Mobile Web Performance Optimization

Deliver a better mobile user experience by improving and optimizing your website – follow these practical steps for cutting-edge application development

S. S. Niranga

PUBLISHING

BIRMINGHAM - MUMBAI

Mobile Web Performance Optimization

First published: December 2015

Production reference: 1161215

Published by Packt Publishing Ltd.
Livery Place
35 Livery Street
Birmingham B3 2PB, UK.

ISBN 978-1-78528-997-2

www.packtpub.com

Credits

Author
S. S. Niranga

Reviewer
Ankit Aggarwal

Commissioning Editor
Veena Pagare

Acquisition Editor
Sonali Vernekar

Content Development Editor
Pooja Mhapsekar

Technical Editor
Mohita Vyas

Copy Editor
Imon Biswas

Project Coordinator
Francina Pinto

Proofreader
Safis Editing

Indexer
Rekha Nair

Graphics
Abhinash Sahu

Production Coordinator
Melwyn Dsa

Cover Work
Melwyn Dsa

About the Author

S. S. Niranga is a senior tech lead at Netstarter Pvt Ltd, and he has more than 9 years of experience as a software engineer and a web developer. During this period, he has built more than 300 websites including numerous e-commerce websites, such as JAX Tyres, ActiveSkin, Athlete foots, JVC, Pegasus, and the world's first Magento 2 website, Venroy. Also, he is an active developer on Upwork as well.

Niranga is a certified Magento frontend developer, a Microsoft technical specialist, and a scrum master. Currently, he is pursuing a master's degree in IT at the Sri Lanka Institute of Information Technology.

Niranga has done a few Tech Talk sessions regarding web optimization and e-commerce. This is his first effort as an author.

When I was about to give up, you were always there to support me and cheer me up. Your help and encouragement were what made this book possible. Thank you Iresha Wijethunga, for your understanding and love during the past few years.

Also, I would like to thank Asanka Mawilmada for his support and guidance over the past few years. You are an awesome friend!!!

About the Reviewer

Ankit Aggarwal has been fascinated with science and technology since childhood. He likes to experiment and learn about new things. He is a software engineer and researcher by profession and loves computer science. He wants to solve world problems using technology. His interests range from science, to technology, academic research, music, photography, entrepreneurship, DIY, movies, anime, and much more.

He has been working in networking, distributed systems, pervasive/mobile computing, data science, AI, computer vision, and the list goes on. He is a published author of IEEE Xplore research papers and an active contributor and author on multiple open source projects. He is socially active, blogs occasionally, and maintains his website on `http://ankitaggarwal.me`.

In his free time, he reads, takes part in competitive programming, captures nature with a lens, and watches TV shows, movies, and anime. When he is not doing these things, he can be found jogging at the nearest ground.

www.PacktPub.com

Support files, eBooks, discount offers, and more

For support files and downloads related to your book, please visit www.PacktPub.com.

Did you know that Packt offers eBook versions of every book published, with PDF and ePub files available? You can upgrade to the eBook version at www.PacktPub.com and as a print book customer, you are entitled to a discount on the eBook copy. Get in touch with us at service@packtpub.com for more details.

At www.PacktPub.com, you can also read a collection of free technical articles, sign up for a range of free newsletters and receive exclusive discounts and offers on Packt books and eBooks.

https://www2.packtpub.com/books/subscription/packtlib

Do you need instant solutions to your IT questions? PacktLib is Packt's online digital book library. Here, you can search, access, and read Packt's entire library of books.

Why subscribe?

- Fully searchable across every book published by Packt
- Copy and paste, print, and bookmark content
- On demand and accessible via a web browser

Free access for Packt account holders

If you have an account with Packt at www.PacktPub.com, you can use this to access PacktLib today and view 9 entirely free books. Simply use your login credentials for immediate access.

Table of Contents

Preface

This book is for anyone who has basic knowledge of web development and who wants to enhance their knowledge on mobile website performance optimization. By reading this book, a user will get to know how to measure their website's performance, the tools they can use to debug and monitor their website, and the tips and tricks to optimize their website.

What this book covers

Chapter 1, *Pillars of Mobile Web Performance Optimization*, discusses mobile history and why mobile web optimization is necessary. Also, we will talk about the three main pillars that are important in the mobile world, and also discuss the major browsers and popular OSes in the market.

Chapter 2, *Mobile Web Optimization Essentials*, explains the importance of reducing HTTP requests and enabling Gzip on the server and its benefits. We will discuss the importance of image optimization and the tools we can use. Also, we will see content management, the importance of UX, and how it affects a mobile site.

Chapter 3, *How to Optimize Yours Mobile Website*, discusses HTML5 and CSS3 and how to use their features for performance optimization. We will especially talk about the importance of hardware acceleration and GPU, CSS3 animations versus JavaScript animations, and how to use iconic fonts instead of images. After that, we will see how to use media queries and display none in CSS. We will also explore CSS preprocessors and the importance of minifying CSS and JS.

Chapter 4, *Caching and Optimizing*, shows how the caching mechanism works. After that, we will see how a developer should call JavaScript and CSS files and why we should avoid empty source and link attributes. Then we will have a brief introduction to CSS and JavaScript frameworks. The later part of the chapter explains how we can optimize JavaScript to gain performance and the importance of reducing DOM elements.

Chapter 5, Monitoring and Debugging Our Website, demonstrates how to use profiling tools such as GPU Overdraw Walkthrough and GPU Rendering Walkthrough. After that, we will see the features of browser's DevTools and how we can remote debug our website using devices actually connecting to our PC. Also, we will discuss the Firefox, Safari, and IE developer toolbar and how we can use them for debugging. In the later part of the chapter, we will go through the Google emulator and see how we can use it as a testing environment. Finally, we will see how to get a performance score and rating for our website using Google PageSpeed and YSlow.

Chapter 6, Managing Third-Party Components, teaches you how to check 404 errors in our website, why it is important to eliminate 404 errors, and how we can do that. Not only 404, but we should also learn and understand 300, 400, and 500 error messages as well. Then, we will discuss CDN networks and the benefits that we can get using a CDN network. Then we will cover how to open and close connection works and the importance of offloading to Wi-Fi. After that, we will discuss screen rotation and how we can use it to optimize our website. Finally, we will see Adobe Flash and you are recommended not to use it.

Chapter 7, Tips and Tricks, discusses why we should build for performance and how we can convince our clients to approve a budget for performance. Also, in the chapter, we will see what the limitations of our design tools are and how we can get the best out of them. Finally, we will discuss the New Relic mobile app, a tool that we can use to monitor our application's performance. The tool generates a very detailed report, which helps developers in many aspects.

What you need for this book

The software used in this book are as follows:

Software required	Free/Proprietary	Download links to the software
Tiny png	Free	https://tinypng.com/
ImageOptim	Free	https://imageoptim.com/
Kraken	Free	https://kraken.io/
Font Awesome Icons	Free	https://fortawesome.github.io/ Font-Awesome/icons/
IcoMoon Icons	Free	https://icomoon.io/
SASS	Free	http://sass-lang.com/
LESS	Free	http://lesscss.org/
GPU Overdraw Walkthrough	Free	

Software required	Free/Proprietary	Download links to the software
Browser DevTools Performance Tools	Free	
Firefox Developer Tools	Free	
IE 11 Developer Tools	Free	
Safari Developer toolbar	Free	
YSlow	Free	`http://yslow.org/`
Web page test	Free	`www.webpagetest.org`

Who this book is for

This book is for anyone who has basic knowledge of web development and who wants to enhance their knowledge of mobile website performance optimization.

Conventions

In this book, you will find a number of text styles that distinguish between different kinds of information. Here are some examples of these styles and an explanation of their meaning.

Code words in text, database table names, folder names, filenames, file extensions, pathnames, dummy URLs, user input, and Twitter handles are shown as follows: "In Apache, you can add the following code to the .htaccess file."

A block of code is set as follows:

```
# compress text, html, javascript, css, xml:
AddOutputFilterByType DEFLATE text/plain
AddOutputFilterByType DEFLATE text/html
ddOutputFilterByType DEFLATE text/css
AddOutputFilterByType DEFLATE application/javascript
AddOutputFilterByType DEFLATE application/x-javascript
AddOutputFilterByType DEFLATE text/xml
AddOutputFilterByType DEFLATE application/xml
AddOutputFilterByType DEFLATE application/xhtml+xml
AddOutputFilterByType DEFLATE application/rss+xml

# Or, compress certain file types by extension:
<files *.html>
SetOutputFilter DEFLATE
</files>
```

New terms and **important words** are shown in bold. Words that you see on the screen, for example, in menus or dialog boxes, appear in the text like this: "Go to the taskbar and click on **Server Manager**."

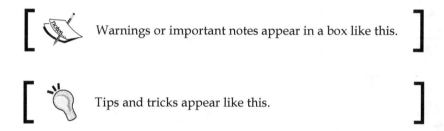

> Warnings or important notes appear in a box like this.

> Tips and tricks appear like this.

Reader feedback

Feedback from our readers is always welcome. Let us know what you think about this book—what you liked or disliked. Reader feedback is important for us as it helps us develop titles that you will really get the most out of.

To send us general feedback, simply e-mail feedback@packtpub.com, and mention the book's title in the subject of your message.

If there is a topic that you have expertise in and you are interested in either writing or contributing to a book, see our author guide at www.packtpub.com/authors.

Customer support

Now that you are the proud owner of a Packt book, we have a number of things to help you to get the most from your purchase.

Downloading the color images of this book

We also provide you with a PDF file that has color images of the screenshots/ diagrams used in this book. The color images will help you better understand the changes in the output. You can download this file from https://www.packtpub.com/sites/default/files/downloads/MobileWebPerformanceOptimization_ColoredImages.pdf.

Errata

Although we have taken every care to ensure the accuracy of our content, mistakes do happen. If you find a mistake in one of our books—maybe a mistake in the text or the code—we would be grateful if you could report this to us. By doing so, you can save other readers from frustration and help us improve subsequent versions of this book. If you find any errata, please report them by visiting http://www.packtpub. com/submit-errata, selecting your book, clicking on the **Errata Submission Form** link, and entering the details of your errata. Once your errata are verified, your submission will be accepted and the errata will be uploaded to our website or added to any list of existing errata under the Errata section of that title.

To view the previously submitted errata, go to https://www.packtpub.com/books/ content/support and enter the name of the book in the search field. The required information will appear under the **Errata** section.

Piracy

Piracy of copyrighted material on the Internet is an ongoing problem across all media. At Packt, we take the protection of our copyright and licenses very seriously. If you come across any illegal copies of our works in any form on the Internet, please provide us with the location address or website name immediately so that we can pursue a remedy.

Please contact us at copyright@packtpub.com with a link to the suspected pirated material.

We appreciate your help in protecting our authors and our ability to bring you valuable content.

Questions

If you have a problem with any aspect of this book, you can contact us at questions@packtpub.com, and we will do our best to address the problem.

1
Pillars of Mobile Web Performance Optimization

If you are into mobile web or application development, it's essential to learn about the basics of mobile and how it has evolved over the last few decades. By learning about these topics, you will gain basic knowledge about mobiles, which will help you to understand the concepts that we are going to discuss in later chapters. Also, in the context of mobile web optimization, you can't ever forget the three main constraints that mobile devices have, and you are going to learn about these three constraints later.

In this chapter, we will discuss the following topics:

- Brief history of mobile development
- Three main pillars
- Available browsers
- Mobile OS

A brief history of mobile development

It is said that:

> *"The Apollo 11 mission's computers were less powerful than today's mobile phones."*

In 1970, a year after the human race set foot on the moon, Martin Cooper of Motorola conceived the idea of the first handheld mobile phone. Since then, the mobile phone has evolved at a rapid rate, and evidence showed that it's not going to stop any sooner. It's difficult to imagine how we made such an advance in mobile technology in such a short period, and reached a point where today, most of us use mobile devices to complete many activities in our day-to-day life.

I still remember the day I bought my first mobile phone. It didn't have any fancy stuff that you find in today's mobile phones. The only advanced feature that it had was the **Short Message Service (SMS)**. It didn't have a camera, all the applications were pre-installed, the user couldn't install any applications, and there was no Internet browsing.

However, nowadays, we use mobile devices for many things because communication through a mobile device is faster, cheaper, and can connect to anyone from anywhere. According to surveys, the number of active mobile devices and human beings crossed over somewhere around the 7.19 billion mark. It means that each and every person in the world most likely has a mobile device. Because of this large consumer group, many organizations and consumers invested and made their marketing campaigns to cater to mobile users and as a result, each and every day thousands of new mobile applications and mobile websites have been introduced into the market.

However, today mobile applications and mobile websites have a fatal flow. Websites' sizes are getting bigger at an alarming rate, and we are quickly heading towards the wrong way. We never notice it as it happens, and when we do, it's often too late.

I had the privilege to work with excellent internal developers to complete a website a couple of months back, and our initial goal was to build the website in such a manner that it loads at top speed. Although we planned everything upfront to achieve our goal, we made a fundamental mistake. When we saw the designs, it was already approved by the top management and we never saw the designs upfront before they were sent to the client. Then we got the internal deadline defined by the management, and it was too tight. Then, *Make it fast* turned into *Make it work* and we thought we can make it faster later; of course that later never came.

After a couple of months' hard work, we managed to launch the website, but it was a disaster. The site looked great in frontend, but it took more than 20 seconds to load the home page. The website was responsive, and when we came into a mobile breakpoint, it loaded a lot of unwanted elements that shouldn't be there. Once we saw this flaw, we had to work very hard even at night to tweak the website, and after a massive effort, we managed to load the website within 7 seconds.

That day we promised ourselves to check and plan everything upfront, and never leave anything behind to damage the site's performance. So, in this book, I am going to discuss a few tips, tricks, and tools that I have learned in the past couple of years. I hope it will help you to improve your website's loading time by at least a couple of seconds.

Remember, many studies and surveys have shown how a website's performance has a direct impact on the user's interaction with the website. I've listed a few of these as follows:

- 4 percent of mobile phone users visiting a mobile site will abandon a site that takes longer than five seconds to load. (Source of this information: `https://blog.kissmetrics.com/loading-time/`.)

- Every additional second added to the site's load time results in a 7 percent loss in conversions.

- Similarly, shaving two seconds off of Mozilla's landing pages led to a 15.4 percent increase in conversions, which meant 60 million more downloads per year. This is just a two-second difference, so any impact that you can make will have the potential to improve your business. (Source of this information: `http://www.yottaa.com/blog/application-optimization/marketing-web-performance-101-how-site-speed-impacts-your-metrics-`.)

- 46 percent of the people who abandon their shopping carts cite slow website speeds to be the reason to do so.

- 79 percent of shoppers who are dissatisfied with the website's performance are less likely to buy from the same site again. (Source of this information: `http://conversionxl.com/11-low-hanging-fruits-for-increasing-website-speed-and-conversions/`.)

Three main pillars

It's true that the mobile phone has come a long way since 1970. Today, we use the mobile phone for navigation, communication, and entertainment. We even use it as an electronic valet. Although, we use mobile devices for hundreds of different tasks, it has a few limitations. Mobile devices have a limited screen size, so whatever you do, you have to build your application or website in a way that it fits into that limited space. Also, if your device is truly a portable device, you will be able to carry it around. As a result, mobile manufacturers have to compact everything into a smaller size.

Finally, when you develop mobile websites or applications, you have to consider the features that mobile users actually expect from their devices, which are as follows:

- Speed
  - 64 percent of mobile users expect pages or apps to load in less than 4 seconds. (Source: `https://econsultancy.com/blog/10936-site-speed-case-studies-tips-and-tools-for-improving-your-conversion-rate/`)

- Battery
  - ° Better battery life (6.1 people out of 10 are satisfied by this)
  - ° 72 percent of the people rate their phone as very good or excellent
  - ° Users don't want to exceed their datacap
- Reasonable data usage
  - ° Users don't want to exceed their datacap

The following chart explains the significance of the three main features for any mobile device:

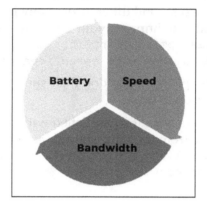

Battery

Out of these three factors, our primary focus will always be on the mobile's battery because unlike personal computers, mobile phones do not have an unlimited power source. Most of the time, users have to charge their mobile phones daily. It's a known fact that when you turn on the mobile's Wi-Fi, 3G, or 4G data connections, your battery starts to drain. This is because once you turn on Wi-Fi, 3G, or 4G data connections, your mobile phone begins to exchange data, thereby consuming more power.

No one will want to visit your website or use your application if they feel that your website or application is draining their battery. So, you should always optimize your website or application in such a way that it uses minimum power. To do that, you need to have a better understanding about energy consumption in mobile phones.

How a 3G wireless state machine works

In the following diagram, you can see how a 3G wireless state machine works:

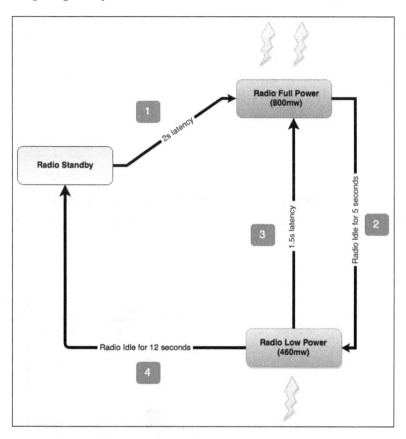

Learning about this will give you an idea about how mobile devices consume power when they start to exchange data using 3G:

- As the preceding diagram indicates, the mobile device is initially on standby mode (**A**).

- As number **1** indicates, when we start the 3G connection, it will have a 2 second latency. The mobile device will push to its maximum state and begin to consume maximum power (**A** to **B**).

- The mobile device will continue to be in this state as long as it receives data. As long as it is connected to your website or app it will use battery power at a higher rate (**B**).

- If your device didn't get data from the website or app, it will keep the connection for another 5 seconds, after which it will go to a lower power state (460mw) – number **2** (**B** to **C**).

- If the device makes a connection with the website or app at this stage again, it will take 1.5 seconds to go to a higher level of power consumption – (**C** to **B**).

- If the device didn't succeed to make the connection with the website or app (**C** state), it would remain there for another 12 seconds and after that it will come to a standby mode (**C** to **A**).

Assuming that you have gained the basic knowledge about how 3G wireless state machine works, let's now see how the 4G **Long-Term Evolution** (**LTE**) wireless state machine works.

How a 4G LTE wireless state machine works

In the following image, you can see how a 4G LTE wireless state machine works:

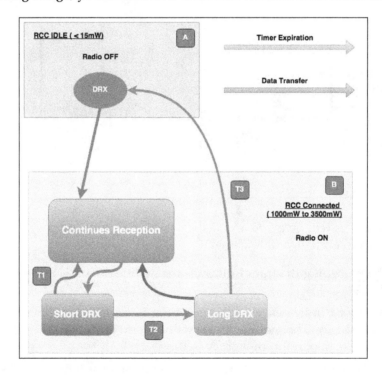

Following is a description of the preceding figure:

- In the idle state, the radio is off and uses low power (<15mW) (**A**)
- In the connected state, the radio is on and uses high power (1000 − 3500 mW) while it either transmits data or waits for data (**B**)
- When data is received, the machine goes to the **Short DRX** mode (**T1**)
- If there is no data, it switches to the **Long DRX** mode (**T2**)
- In the **Long DRX** state, the radio prepares to switch to the idle state but it's still using high power and waiting for data
- If more data arrives, then the radio returns to the continuous transmission state
- If it does not receive any more data, it switches to the low power (<15mW) idle state and switches off (**T3**)

In the context of both 3G and 4G data connections, as long as we keep the connection open to receive data, the mobile goes to full power and starts to consume more battery power.

For example, take a person who is supposed to bring goods from a supermarket and they use their vehicles to bring them in. If they didn't get any request to bring anything, then they are in an idle state, and do not waste any energy. Suddenly, someone comes up to them and asks them to bring a pack of sugar from the supermarket. So, now they have to start the car and go to the supermarket to bring that item. In this particular event, they have consumed some energy. Once they deliver the goods, the same person asks them to bring another thing from the supermarket. Now, they again have to proceed with the same routine, which again leads to consumption of energy.

Imagine a situation where they have to go through the same routine a couple of times. Apparently, it wastes both the person's time and money, and it could have been easily avoided if the task was planned properly and all the items were requested at the same time.

The same situation holds true for mobiles as well. So, keep in mind that when you develop a website or app, you have to think about these little details to build an optimized website or app.

Opening and closing connections

As you can see, mobile devices drain more battery power as long as they go to the 3G and 4G mode. So, in order to minimize power consumption, the developer should always remember to close the connection as soon as possible. The main reason for this is that the device goes to higher state, stays there for a longer period, leaves a connection open long after data has been transmitted, and the files are requested by the users in lengthy stretches of time.

However, as developers, we can minimize this issue by taking the following measures:

- Downloading the content as quickly as possible
- Grouping the TCP packets together when opening a connection
- Prefetching the content
- Closing the connection quickly after data is transmitted

By following these methods the developer can reduce energy consumption and network latency, and can load the website or application faster, which will lead to happier users. The following image shows the issue of a connection that was closed inefficiently. In the **Bursts** row of this image, the bursts (in green and red) between 10 and 40 seconds on the timeline were used to download data. While the burst (in blue) around the 80 second mark represents a request to close the earlier connection. Notice that the radio, represented on the **RRC States** row below the Bursts row, is turned back on just to close the connection. This can be seen at the 80 second mark below the narrow burst (in blue). The **RRC States** row also shows that the radio is not just simply turned on and off. It remains on for a set period, including time in a high energy state, then low energy, and all the related tail time (wasted energy) before it turns off.

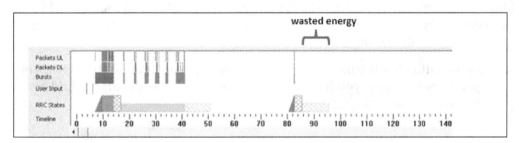

Speed

When a customer enters the supermarket, they always seek and expect a fast and friendly service from the staff. If they got a good service from the store, chances are really high that they shared that experience with their friends and family, which will lead to more customers going to the supermarket. Also, if they have to wait for a long period in a queue to get clarification for something or if the support staff didn't offer a friendly service to them, they will share these bad experience with their friends. This will damage the store's reputation, which is very hard to restore. This is why many companies nowadays spend millions and billions of dollars on customer services. They will always try to keep the customer happy.

This is true for your website or app as well. If you want to attract more clients or users to your product, you should give 100 percent speed to the visitors all the time. This is why you should always provide an optimized website or app to the client. So, once they visit your application, they will get a smooth and fast experience that will generate more revenue. Also, unlike desktop users, mobile users don't stick to one place for a long period of time. They won't wait until your website gets loaded, they will just ignore your website and will visit your competitor's website. This is why speed matters.

According to researchers, a fast browsing experience will increase the following:

- **Minimum bounce rates**: The percentage of visitors to a particular website who navigate away from the site after viewing only one page
- **Order size :** Customers will order more products from your website
- **Customer satisfaction:** Customer will love to browser your website
- **SEO rankings:** Your website will get a better ranking from the search engines

In Google site rankings, loading time is given more weight (source: `http://googlewebmastercentral.blogspot.com/2010/04/using-site-speed-in-web-search-ranking.html`). If you expect your website to have an increase in the rankings, you should consider your application's performance, and this is why optimization is necessary.

Bandwidth

When I was in school, I used a dial-up connection to connect to the Internet. It was a 56 K connection, and I still remember that it took ages to download a 3 MB file. Since then, Internet service providers have come a long way, and now many of us can download a complete movie in less than 5 minutes.

Although now we have faster connections, in many countries, the Internet is still not that cheap. Users have to pay a premium to get a connection, and they have to pay a monthly payment to sustain it.

In many countries, now it's easy to get a connection; with mobiles, it's just following a few steps and within 30 minutes, most of us can obtain a 3G or 4G connection. However, many of these packages have a bandwidth constraint. For an example, users will get faster connection with X GB but, once the user exceeds it, they have to pay extra.

So, when developing a website or app, you should always keep in mind that consuming their bandwidth immensely will result in them ignoring your website or app and moving on to an alternate website.

Managing the bandwidth is not that difficult, go through a few simple steps and you will be able to save a few extra MBs of your site, which will help you to increase revenue.

Available browsers

When building a mobile website, you should always identify the sort of browser that the end user uses. These days, looking at analytics data, the developer can easily find out what type of customers they have and their needs. By having those data upfront, the developer can easily use browser features more effectively. In this section, we will discuss some of the major browsers available in the market.

Browser	Sessions	% Sessions
1. Safari	7,142	50.56%
2. Chrome	3,277	23.20%
3. Internet Explorer	2,349	16.63%
4. Firefox	761	5.39%
5. Opera	276	1.95%
6. YaBrowser	174	1.23%
7. Android Browser	51	0.36%
8. BlackBerry	24	0.17%
9. (not set)	22	0.16%
10. Safari (in-app)	21	0.15%

Analytic data

You can find the top browser share trend of mobile/tablet in the following screenshot and the source to this data is `https://www.netmarketshare.com/browser-market-share.aspx?qprid=1&qpcustomb=1`:

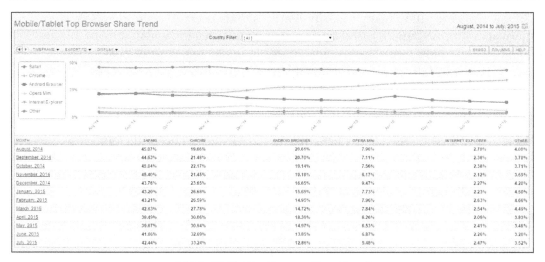

MONTH	SAFARI	CHROME	ANDROID BROWSER	OPERA MINI	INTERNET EXPLORER	OTHER
August, 2014	45.07%	19.66%	20.61%	7.96%	2.70%	4.00%
September, 2014	44.63%	21.46%	20.70%	7.11%	2.38%	3.70%
October, 2014	45.04%	22.17%	19.14%	7.56%	2.38%	3.71%
November, 2014	45.40%	21.45%	19.18%	8.17%	2.12%	3.69%
December, 2014	43.76%	23.65%	16.65%	9.47%	2.27%	4.20%
January, 2015	43.20%	26.68%	15.65%	7.73%	2.23%	4.50%
February, 2015	43.21%	26.59%	14.95%	7.96%	2.63%	4.66%
March, 2015	42.63%	27.78%	14.72%	7.84%	2.54%	4.49%
April, 2015	39.49%	30.06%	18.30%	6.26%	2.06%	3.83%
May, 2015	39.67%	30.94%	14.97%	8.53%	2.41%	3.48%
June, 2015	41.66%	32.09%	13.85%	6.87%	2.26%	3.26%
July, 2015	42.44%	33.24%	12.86%	5.48%	2.47%	3.52%

Mobile/Tablet Top Browser Share Trend

Safari

Known for its natural ease of use, Safari is Apple's lightweight and smooth web browser. Since 2007, Safari has become the most favorite browser in the mobile world. The Safari browser is quick and simple to use. However, it does not have the customization option that a large number of clients look for in a browser nowadays.

Each and every browser has some unique features, and Safari does too. Safari is amazingly quick. It takes less than two seconds for the program to load and even less time to navigate interfaces on the site. Safari offers various features to users, such as tabs, spellcheck, and a secret key administrator. However, customizing the browser is a bit difficult. Also, Safari additionally needs parental and zoom controls.

One other main benefit that Safari has is its security features. The browser provides security from a wide range of malware and phishing sites.

Chrome

Since 2008, after the first release, Chrome has gradually gained the largest market share in global Internet usage. The browser upholds Google's reputation for innovation and industry dominance. In the beginning, the Chrome browser got many ideas from other browsers, but now, other browsers are inspired by Chrome.

Google Chrome's best features are simplicity and speed, which is better than other browsers. The browser has earned many awards for its minimum loading time and seamless navigation.

Google Chrome offers many security features to its users to keep them safe from malware and phishing. Its auto-update feature ensures the installation of all the latest security fixes with ease. When a user navigates to a website that contains malware or phishing, this browser displays a warning.

Internet Explorer

When compared to other browsers, Internet Explorer has been the longest in the run. It was the most popular browser in the past, but it was suppressed by others because of its lack of security and features. However, recently Microsoft has placed a heavy focus on enhancing the security and features to give more options to users.

Internet Explorer may be not the fastest browser on the market, but Internet Explorer has many new features such as tabbed browsing and most visited sites based on browsing history that works great with the touchscreen. Also, the browser provides a variety of add-ons for a fully customized browsing experience. The add-ons are categorized into four groups such as accelerators, search providers, Web Slices, and toolbars. In each category, the user can find many downloads, and most of them are free.

Internet Explorer has had an awful run in the past; many hackers attacked Windows OS because of its larger market share, and it has affected Internet Explorer very badly. However, the latest version of the browser had included effective updates and patches to reduce those loopholes, and the browser provides a very user-friendly interface.

Firefox

Mozilla developers are always offering products that represent the open web concept. They always try to keep the standards of their product and versions. Firefox, Mozilla's web browser, has always upheld these standards and has given a true mobile experience to its users.

All the features in Firefox are fairly standard, and their security features always keep the users in a safe place. Also, the user can easily sync this mobile browser with their desktop versions of Firefox and the home panel for the app is customizable with the user's choice. Adding to this, the swiping gestures and simple interfaces that Firefox has work well in smartphones and tablets, which gives a very pleasant experience to its users.

Opera mini

As I have mentioned in the previous section, if you have a limited mobile data package, every byte you download is really important. If that person is you and, if you don't care about the fancy features that conventional browsers offer, you should go with Opera Mini.

In the context of Opera Mini, the browser has a best image compression mechanism, and auto-play videos are disabled by default. This will save you 90 percent data compared to other browsers.

On the surface, Opera mini doesn't have a lot of variations compared to Opera's main version or any other browser. However, when you try to download an image-heavy site such as Facebook or Tumblr, you will notice the difference. The images that you are downloading from these sites look blurry compared to other browsers, but you will save a massive amount of mobile data.

Mobile OS

Choosing a mobile OS is not an easy task. The mobile world is divided into Google Android, Apple iOS, Windows and a few up and coming players. The OS you choose will define the kind of app or website that you are going to build. The good news is when it comes to the mobile web, there isn't much difference from those players.

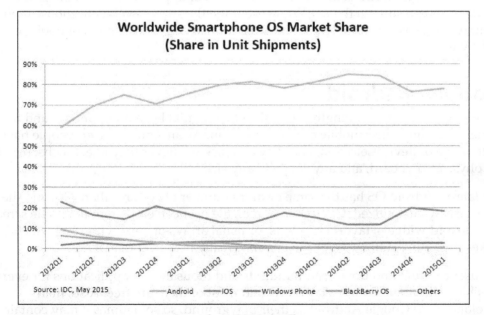

Smartphone OS Market Share

However, the difference between mobile browsers always comes down to three factors: hardware, application, and customizability. At the moment, Google Android has the highest market share, iOS has the most popular apps, while Windows Phone 8 and Blackberry OS 10 lag behind.

Period	Android	iOS	Windows Phone	BlackBerry OS	Others
Q1 2015	78.0%	18.3%	2.7%	0.3%	0.7%
Q1 2014	81.2%	15.2%	2.5%	0.5%	0.7%
Q1 2013	75.5%	16.9%	3.2%	2.9%	1.5%
Q1 2012	59.2%	22.9%	2.0%	6.3%	9.5%
Source: IDC, May 2015					

Apple iOS

Style and simplicity are the main indicators of iOS, which comes with iPhones and tabs. iOS has a very simple, logical, and consistent design throughout the OS, and the home screen contains a grid of movable icons. The built-in applications of iOS are well designed and user-friendly, the new iOS version includes Facebook and Twitter integration, built-in video chat, and the Passbook virtual wallet.

The best strength of iOS is the massive collection of apps and most probably it has the best app store in the market. Most of the time app developers choose their primary target as iOS, and Apple offers the industry's best collection of books, music and TV to its users.

Google Android

Google Android's best strength is that the OS is available on more phones and more carriers than any other mobile operating system. As an Android user, people have a wide range of device selection, and they can pick whatever they like, touchscreen to the physical keyboard, and any shape to any size.

The latest Android OS has the same or more number of features than iOS, and the OS can be easily customized. Compared to iOS, the user can easily edit the home screen and can add widgets, favorite contacts, or usual arrays of apps with ease, which makes the Android home screen experience really powerful than iOS.

The user can find and download hundreds and thousands of applications for every possible scenario in Google Play Store, and most of them are free. Also, many developers use Google Android as their playground, so sometimes it may contain some security issues as well.

Microsoft Windows Phone 8

Microsoft's mobile OS has the balance between iOS' simplicity and Android's customizability. The main feature of this OS is live tiles, which are preprogrammed squares that the user can easily rearrange it as they desire. Windows Phone 8 has inbuilt Facebook and Twitter and works brilliantly with Microsoft Exchange, MS Office, and XBOX live to game.

However, compared to iOS and Android, Windows mobile OS gets a lower score. This is because of limited hardware options and limited applications availability than iOS and Android. The operating system uses the Bing search engine, which scores well behind Google's on accuracy and features.

BlackBerry 10 OS

With Blackberry's new OS, the user can access a universal inbox that has all the e-mails and social-network messages with ease. It has an efficient and clean interface in the OS, mainly focusing on communication and messaging, and it already has more than 100,000 apps. However, some of the popular and useful apps from iOS and Android are still missing.

The OS home screen, customizable to some extent, is similar to Android and Windows mobile OS, but the user cannot add widgets or contacts as icons the way the user can on Windows and Android. Blackberry has the best touch keyboard in the market, and their design is much easier to type on.

Blackberry has a Web kit-based web browser, and it uses technology from Torch Mobile. The browser has a private browsing mode, desktop mode, and it uses Bing as the default search engine, but the user can change it.

Summary

In this chapter, we have discussed the history of mobiles and why mobile web optimization is necessary. Also, we have discussed the three main pillars that are important in the mobile world, and you have seen how 3G and 4G data connections can drain the user's battery. After this, we went through the major browsers and popular OSs in the market, and we have discussed negatives and positives in those.

In the next chapter, we will take a look at the essential components in mobile web optimization. The chapter will help you learn about the differences between mobile sites and responsive websites, and you will also learn some of the image optimization tools and more.

2
Mobile Web Optimization Essentials

By now you know the limitations of the mobile devices and why Mobile Web Optimization is necessary. In this chapter, I will share with you some of my experiences that I have learned from in the last couple of years. The majority of these techniques are fairly easy to implement, but the outcome that they produce is huge. So, I encourage you to go through each section thoroughly because they hold the key to revealing the secrets of Mobile Web Optimization world.

Also, I assume you at least have a basic knowledge of web design and frontend web development to understand this chapter properly. However, if you are new to the game, don't worry; I can assure you going through the following sections is not going to be a waste of time, and you will get many things out of them that you are going to remember for a long time.

In this chapter, we are going to discuss the following sections:

- Mobile-only websites versus responsive websites
- How to reduce HTTP requests
- Image size matters
- Unnecessary contents
- Why design and UX are important

Mobile-only websites versus responsive websites

A couple of years ago I was given the opportunity to build a website for an insurance company. The website was modern, and it had many features. Even a user was able to customize and purchase an insurance policy by spending a couple of minutes browsing the website. After launching the website, the client realized that he had an excellent opportunity to enter the mobile market because back then, mobile browsing was in its early stages. So, the following month he contacted us and asked us to build a mobile version of the website.

When we read the requirement, we realized that the mobile website had a similar functionality; only the UI was going to be different. So, we used the same codebase that the desktop version had, and we created a new website using the existing code. This was a popular method at that time, and the mobile website had a different URL. Once the user entered the original website using a mobile device, our script detected the device and redirected the user to the mobile website. This is how a Mobile-only website works. However, this method has some flaws.

To avoid these flaws, we now use another method to build the mobile websites. It is not another website, we use CSS media quarries and give a different look and feel to the website by detecting the device screen size. Once we do this, the desktop, tablets, and mobiles will have the same codebase but a different look. This is what we call *Responsive web design*.

In the following code, you can see how it is possible to have different backgrounds for various screen sizes using the CSS media quarries:

```
@media screen and (max-width: 320px) {
    body {
        background-color: Yellow;
    }
}
@media screen and(min-width: 321px) and (max-width: 768px) {
    body {
        background-color: Pink;
    }
}
@media screen and (min-width: 769px) and (max-width: 1024px) {
    body {
        background-color:Green;
    }
}
```

If we compare the Mobile-only websites and responsive websites, we will find that both of them have positive as well as negative qualities. However, if you are going after 100 percent performance of your mobile website and you don't want to worry about any other stuff, my recommendation will be that you go for the Mobile-only website because you will have a separate code for the mobile device, and you can only use mobile-related content. However, as mentioned earlier, it has some negative impacts as well.

In the following table, you will see the comparison between Mobile-Only site versus Responsive Design.

	Mobile-only website	Responsive website
Domain Protection	Mobile-only sites will have a different URL (`m.domain.com`) that will hurt organic search traffic. Developers have to manage two code bases.	Responsive websites use the same domain, and only the backend code will differ. Many search engines show that this is the best method for SEO.
Link Equity	Mobile-only websites use a separate domain, which does not count as a primary site URL, and is not useful when searching online.	Responsive websites have the same URL, which counts as the primary site URL. This can be whatever the device used to browse, which would be the best option for searching online.
Rendering Experience	The Mobile-only website uses a different code base that is only allocated to mobile browsing. Using optimization techniques and new browser features, the developers can provide an optimized solution to the end user.	This is a flexible solution, but sometimes mobile devices will have some unwanted elements related to a desktop site that will have some impact on performance.
Future-Ready	The developers have to maintain two code bases that are expensive and difficult to maintain	This technology is more future ready because developers can add or change the layout easily by simply manipulating CSS.

HTTP requests

Do you remember the example of the supermarket that was given in *Chapter 1, Pillars of Mobile Web Performance Optimization*? In that instance, I explained how inefficient we are going to be if we make round trips to buy each and every item from the shop. To overcome this issue, most of the time, we make a list before going shopping, and we add all those items into our shopping cart before making the payment.

We can apply this technique to our website as well. According to researchers (source: `https://developer.yahoo.com/performance/rules.html`), the end users spend 80 percent of their response time on downloading frontend contents. Most of this time is associated with downloading images, scripts, and style sheets that are essential to the rendering of the page. However, if we can reduce these components that are getting downloaded, then we can reduce the HTTP requests.

When doing a mobile web optimization, reducing the number of HTTP requests in your website is the best place to begin. According to researchers, 40-60 percent of daily visitors come to a website with an empty cache and they probably are new visitors. So, making the mobile site fast for those visitors is key to a better user experience.

By simplifying the design of the page, we can easily reduce the number of HTTP requests of the page, but we cannot always have a simplified design because sometimes, we need content-rich websites. In this section, we are going to discuss a few techniques for reducing the number of HTTP requests.

In the following screenshot, you can see some of the content that is downloaded when we browse a mobile website:

Name	Method	Status
styles.css?v=2081	GET	200
boilerplate.css?v=2081	GET	200
owl.carousel.css?v=2081	GET	200
owl.theme.css?v=2081	GET	200
jquery-ui.css?v=2081	GET	200
common.css?v=2081	GET	200
mainNavigation.css?v=2081	GET	200
checkout.css?v=2081	GET	200
internal.css?v=2081	GET	200
product-details.css?v=2081	GET	200
product-category.css?v=2081	GET	200
jquery.css?v=2081	GET	200
uniform.default.css?v=2081	GET	200
dashboard.css?v=2081	GET	200
dashboard-mobile.css?v=2081	GET	200
widgets.css?v=2081	GET	200
colorbox.css?v=2081	GET	200
jquery-1.8.3.min.js	GET	200
prototype.js	GET	200
ccard.js	GET	200
validation_activeskin.js	GET	200
builder.js	GET	200
effects.js	GET	200
dragdrop.js	GET	200
controls.js	GET	200
slider.js	GET	200
js.js	GET	200
form.js	GET	200
menu.js	GET	200
translate.js	GET	200
cookies.js	GET	200
validation.js	GET	200
modernizr_media_query.js	GET	200
scripts.js?v=2081	GET	200

217 requests | 913 KB transferred | Finish: 12.96 s | DOMContentLoaded: 1.42 s | Load: 10.07 s

Console Search Emulation Rendering

Combined files

By combining files together, the developer can easily reduce the number of HTTP requests that they make. By combining all the scripts and CSS into single style sheets, the developer can easily archive this. These days, many platforms offer this service in their backend as a feature. So, if you are a developer make sure that you use those advanced options as much as you can. However, combining files together is not that easy when the scripts and style sheets vary from page to page. Even so, including this part in your release plan improves the response time.

If you have a website that uses some third-party plugins and components, you will get many style sheets and script files. Take a look at the following example. This is a website that was recently built and it had 15 style sheets and 24 scripts calls:

```
<head>
<link rel="stylesheet" type="text/css" href="http://www.mywebsite.com/
css/styles.css" media="all" />
<link rel="stylesheet" type="text/css" href="http://www.mywebsite.com/
css/common.css" media="all" />
<link rel="stylesheet" type="text/css" href="http://www.mywebsite.com/
css/jquery-ui.css" media="all" />
<link rel="stylesheet" type="text/css" href="http://www.mywebsite.com/
css/owl.carousel.css" media="all" />
<link rel="stylesheet" type="text/css" href="http://www.mywebsite.com/
css/widgets.css" media="all" />
<link rel="stylesheet" type="text/css" href="http://www.mywebsite.com/
css/colorbox.css" media="all" />
<link rel="stylesheet" type="text/css" href="http://www.mywebsite.com/
css/jquery.css" media="all" />
<link rel="stylesheet" type="text/css" href="http://www.mywebsite.com/
css/uniform.default.css" media="all" />
<link rel="stylesheet" type="text/css" href="http://www.mywebsite.com/
css/mainNavigation.css" media="all" />
<link rel="stylesheet" type="text/css" href="http://www.mywebsite.com/
css/checkout.css" media="all" />
<link rel="stylesheet" type="text/css" href="http://www.mywebsite.com/
css/internal.css" media="all" />
<link rel="stylesheet" type="text/css" href="http://www.mywebsite.com/
css/product-details.css" media="all" />
<link rel="stylesheet" type="text/css" href="http://www.mywebsite.com/
css/product-category.css" media="all" />
<link rel="stylesheet" type="text/css" href="http://www.mywebsite.com/
css/dashboard.css" media="all" />
<link rel="stylesheet" type="text/css" href="http://www.mywebsite.com/
css/print.css" media="print" />
<script type="text/javascript" src="http://www.mywebsite.com/js/
jquery-1.8.3.min.js"></script>
```

```
<script type="text/javascript" src="http://www.mywebsite.com/js/
prototype/prototype.js"></script>
<script type="text/javascript" src="http://www.mywebsite.com/js/lib/
ccard.js"></script>
<script type="text/javascript" src="http://www.mywebsite.com/js/
prototype/validation_activeskin.js"></script>
<script type="text/javascript" src="http://www.mywebsite.com/js/
scriptaculous/dragdrop.js"></script>
<script type="text/javascript" src="http://www.mywebsite.com/js/
scriptaculous/controls.js"></script>
<script type="text/javascript" src="http://www.mywebsite.com/js/
scriptaculous/slider.js"></script>
<script type="text/javascript" src="http://www.mywebsite.com/js/
varien/js.js"></script>
<script type="text/javascript" src="http://www.mywebsite.com/js/
varien/form.js"></script>
<script type="text/javascript" src="http://www.mywebsite.com/js/
varien/menu.js"></script>
<script type="text/javascript" src="http://www.mywebsite.com/js/mage/
translate.js"></script>
<script type="text/javascript" src="http://www.mywebsite.com/js/mage/
cookies.js"></script>
<script type="text/javascript" src="http://www.mywebsite.com/js/
appmerce/eway/validation.js"></script>
<script type="text/javascript" src="http://www.mywebsite.com/js/
modernizr/modernizr_media_query.js"></script>
<script type="text/javascript" src="http://www.mywebsite.com/js/
scripts.js"></script>
<script type="text/javascript" src="http://www.mywebsite.com/js/
jquery.uniform.min.js"></script>
<script type="text/javascript" src="http://www.mywebsite.com/js/
jquery.selectbox-0.2.js"></script>
src="http://www.mywebsite.com/js/jquery.touchSwipe.min.js"></script>
<script type="text/javascript" src="http://www.mywebsite.com/js/owl.
carousel.min.js"></script>
<script type="text/javascript" src="http://www.mywebsite.com/js/
product_landing.js"></script>
</head>
```

Using CSS and script margin techniques, the developer was able to reduce it to three style sheets and three scripts, as shown here:

```
<head>
<link rel="stylesheet" type="text/css" href="http://www.mywebsite.com/
css/14837ec795666b5925528c0efc58abcd.css" plugins />
<link rel="stylesheet" type="text/css" href="http://www.mywebsite.com/
css/5037f5bb18003f8450323e3332151234.css" media="print" />
<link rel="stylesheet" type="text/css" href="http://www.mywebsite.com/
css/392070536b949cf42f6724d96f21a2c34.css" media="all" />
<script type="text/javascript" src="http://www.mywebsite.com/js/810d4c
d77db5b6cd2688a4a14e6fqwer.js"></script>
<script type="text/javascript" src="http://www.mywebsite.com/js/88ef80
fd45f12b34285737f3133asdf.js"></script>
<script type="text/javascript" src="http://www.mywebsite.com/js/2d1039
1171b5d322415d3c3c16a1zxcv.js"></script>
</head>
```

However, merging files sometimes can create unpredicted issues such as the following:

- Sometimes, it may create namespace conflicts with other scripts and create unpredictable bugs

- If the developer made a change to one file, it would invalidate the whole combined file and browsers will have to cache it again

- Combined files may become large, and it will take additional time to download

- Combined files lose the benefits of having CDN (we will discuss the CDN later) capability

In the following diagram, you can see how merging files affects HTTP requests.

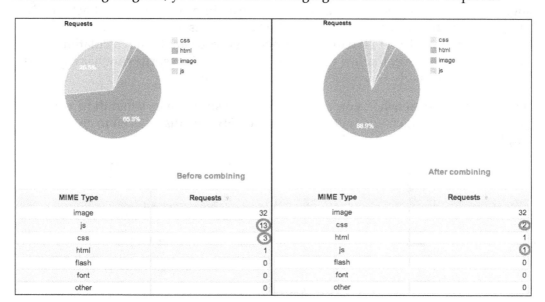

MIME Type	Requests
image	32
js	13
css	3
html	1
flash	0
font	0
other	0

MIME Type	Requests
image	32
css	2
html	1
js	1
flash	0
font	0
other	0

CSS sprites

CSS sprite is not a new technique. It is a well established concept, and many developers use CSS sprites as a common practice. It is true that we do not need to use sprites for every situation, but using sprites will reduce the server load immensely, and it will help us to improve the performance of the page. If you do not have any idea about this technique, now is the time to learn what it is and how it works.

As mentioned earlier, the concept of sprites is not a new invention; it dates back to the mid-1970s. Video game developers used this concept because of the increased complexity of video games at that time. The developers had to deal with a detailed graphic object while keeping the game play as it is, so they used this one large combined image to get the result and the position of the sprite image controlled by the hardware controllers.

Time passed and in the late 2000s, web developers had identified the significance of this method and the web developers started to use this technique on their websites. Creating and using sprites is not a difficult process. The developer has to combine multiple images that have been used throughout the website to make a master image. Then, using the background-position property in CSS, the developer can define the exact position of the image to be displayed.

When the page is loaded, it will load the master image at once rather than loading single images one by one. It might not seem like an improvement, but it actually is. Imagine a situation where a mobile website has millions of page impressions per day. If we can save 10 HTTP requests for one user, you can do the calculation and see for yourself the savings that we can make. This is why CSS sprites are heavily used these days, particularly when we create icons and buttons.

In the following screenshot, you can see one of the sprite images that the developer has created, after which you will find a sample CSS code that is used to create a simple button:

Sample CSS code:

```
.button-clear {
        background: transparent url(sprite.png)0 -210px no-repeat;
    }

.button-clear:hover {
        background-position: 0 -236px;
    }
```

When creating a sprite image, considering the following points will help you to create an optimized CSS sprite:

- Arranging the images horizontally rather than vertically will reduce the file size.
- Combining similar color images into a sprite will help you to reduce the color count.
- Don't leave big gaps in between images. This will not reduce the file size much, but it will help the user agent to decompress the image efficiently.
- Combine PNG and Gif images first.
- You may use a spiriting service that makes the process easy.
- Sprites may be displayed differently in different browsers.
- You have to provide the exact coordinates in the CSS files to define the correct location in the sprite image.
- Combine all the small images to one image so that it will reduce the HTTP request to one.
- Combine cacheable images.

Image maps

Similar to sprites, image maps can be created by combining several images together into a single image. It will keep the same file size, but because of the reduced HTTP requests it will speed up the time taken to load the page. An image map will be ideal for contiguous images and scenarios such as navigation bars. However, defining the coordination of an image map can be a hectic process.

There are two types of image maps available, which are as follows:

- **Client-side**: When a user activates a selected area in an image, the pixel coordination defined to it is identified by the browser and the browser will perform the task allocated to it
- **Server-side**: When a user activates a selected area in an image, the user agent sends that data to the server and the server will perform the task allocated to it

Remove duplicate scripts

Many developers think that the occurrence of duplicate scripts in a web page is very rare but research indicates otherwise. According to Yahoo! (`https://developer.yahoo.com/performance/rules.html`), 20 percent of websites contain a duplicate script. The main reason behind this was that the development team was too large, and there were too many scripts that had been used. When this happens, duplicate scripts decrease mobile performance by creating additional HTTP requests and wasted java script executions. To prevent this, the developers can a use common practice or method, such as a script management module to include scripts.

The usual method to insert a script in a page is to use a script tag on the page:

```
<script type="text/javascript" src="helloworld.js"></script>
```

However, by using an alternative function, the developer can do this easily.

```
<?php insertScript("helloworld.js") ?>
```

Using the preceding method will prevent inclusion of the same script twice on the page and can also be used to do a dependency check and add a version number to script filenames.

Enable Gzip compression

According to the data, Gzip has the capability to reduce the response size by about 70 percent. 90 percent of today's Internet traffic has the capability to support Gzip. Gzip compression is a very simple and efficient method to save bandwidth and speed up the website.

Before we understand Gzip, let's see how our regular browser and server handles a request.

When we request the webpage from the server, it goes through the following process:

1. The browser asks the server to send the `index.html` file.
2. The server receives the request, and searches the `index.html` file.
3. When the server finds the file, it sends the file to the browser.
4. Then, the browser loads the file as it is.

So, this is how the browser and server normally interact with each other. If the server finds a 300 KB page, it sends the file as it is to the browser and the browser will download the 300 KB file and show it to the user.

Consider a situation where the server can send a ZIP file to the browser rather than sending the `index.html` file.

In this case, the process will look like the following:

1. The browser asks the server to send the `index.html` or `index.html.zip` file, if it's available.
2. The server receives the request, and it will search the `index.html` file.
3. When the server finds the file (300 KB), it will zip the file (`index.html.zip`) and send it to the browser (15 KB).
4. The browser receives the `index.html.zip` file, unzips it, and shows it to the user.

To follow the preceding process, the browser and server should have a better understanding about each other, and the agreement has two parts:

- The server will get a message from the browser that it accepts the compressed contents (there are two compression methods, Gzip and deflate)

  `Accept-Encoding: gzip, deflate`

- Then, the server sends the compressed content, if it's available

  `Content-Encoding: gzip`

If the server does not send the compressed content to the requested browser, the browser will take it as a no, and it will start to download the regular version. These days, many browsers have the capability to send the request but our servers are not configured to respond.

Configuring the server is a fairly straightforward process. In Apache, you can add the following code to the `.htaccess` file:

```
# compress text, html, javascript, css, xml:
AddOutputFilterByType DEFLATE text/plain
AddOutputFilterByType DEFLATE text/html
ddOutputFilterByType DEFLATE text/css
AddOutputFilterByType DEFLATE application/javascript
AddOutputFilterByType DEFLATE application/x-javascript
AddOutputFilterByType DEFLATE text/xml
AddOutputFilterByType DEFLATE application/xml
AddOutputFilterByType DEFLATE application/xhtml+xml
AddOutputFilterByType DEFLATE application/rss+xml
```

```
# Or, compress certain file types by extension:
<files *.html>
SetOutputFilter DEFLATE
</files>
```

To enable compression in IIS, you have to first install it on your server. To install static or dynamic compression, use the following steps (`https://www.iis.net/configreference/system.webserver/security/ipsecurity`).

For Windows Server 2012 or Windows Server 2012 R2, use the following steps:

1. Go to the taskbar and click on **Server Manager**.

2. In **Server Manager**, choose the **Manage** menu, and then click on **Add Roles and Features**.

3. In the **Add Roles and Features** wizard, click on **Next**. Select the installation type and click on **Next**. Select the destination server and click on **Next**.

4. On the **Server Roles** page, navigate to **Web Server (IIS) | Web Server | Performance** and select **Static Content Compression** and/or **Dynamic Content Compression**. Click Next.

5. On the **Select features** page, click on **Next**.

6. On the **Confirm installation selections** page, click on **Install**.

7. On the **Results** page, click on **Close**.

For Windows 8 or Windows 8.1, use the following steps:

1. On the **Start** screen, move the pointer all the way to the bottom-left corner, right-click on the **Start** button, and then click on **Control Panel**.

2. In **Control Panel**, click on **Programs and Features**, and then click on **Turn Windows features** on or off.

3. Navigate to **Internet Information Services | World Wide Web Services | Performance Features | Dynamic Content Compression | Static Content Compression**.

4. Click on **OK** and close.

For Windows Server 2008 or Windows Server 2008 R2, use the following steps:

1. On the taskbar, click on **Start**, navigate to **Administrative Tools | Server Manager**.

2. In the **Server Manager** hierarchy pane, navigate to **Roles | Web Server (IIS)**.

3. In the **Web Server (IIS)** pane, scroll to the **Role Services** section, and then click on **Add Role Services**.

4. On the **Select Role Services** page of the **Add Role Services** wizard, select **Dynamic Content Compression** if you want to install dynamic compression or **Static Content Compression** if you want to install static compression, and then click on **Next**.

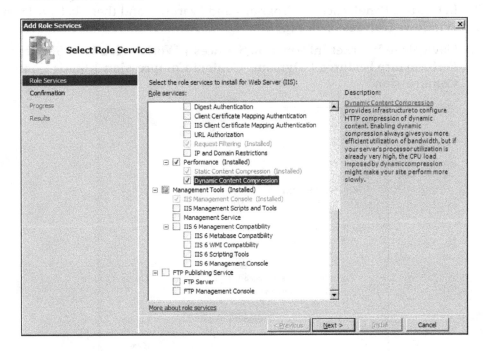

5. On the **Confirm Installation Selections** page, click on **Install**.

6. On the **Results** page, click on **Close**.

For Windows Vista or Windows 7, use the following steps:

1. On the taskbar, click on **Start**, and then click on **Control Panel**.

2. In **Control Panel**, click on **Programs and Features**, and then click on **Turn Windows Features** on or off.

3. Navigate to **Internet Information Services | World Wide Web Services | Performance Features**.

4. Select **Http Compression Dynamic** if you want to install dynamic compression or **Static Content Compression** if you want to install static compression.

5. Click on **OK**.

After you have installed compression you have to enable it for your application or website, and you can do this by using the following steps,

1. Open **Internet Information Services (IIS) Manager**.
2. Next, in the **Connections** pane, go to the connection, site, application, or directory for which you want to enable compression.

3. In the **Home** pane, double-click on **Compression**.

4. In the **Compression** pane, check the boxes to enable static or dynamic compression or remove the check marks to disable static or dynamic compression.

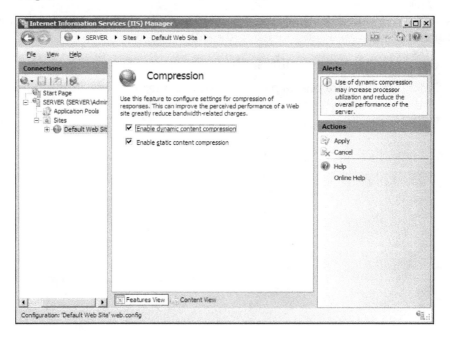

5. Once you have completed the preceding steps, click on **Apply** in the **Actions** pane.

Once you do this, you can check this by using the web developer toolbar on your browser (we will discuss this in a later chapter), or you can use an online Gzip testing tool.

Enabling compress mechanism is one of the fastest ways to improve mobile websites' performance, and it can be done by following some very simple steps. So, enjoy the benefits.

Image size matters

As mentioned in the previous chapter, mobile networks have many limitations compared to a wired connection. So, reducing the file size as much as possible is really essential in mobile web development. According to the data (`https://developer. yahoo.com/performance/rules.html`), 70-80 percent of sites' bandwidth is consumed by the images. Therefore, delivering smaller file size images with acceptable quality to the mobile will always provide a better outcome.

When it comes to image optimization, there are two key factors that you need to always keep in mind. They are as follows:

- Resize your images to correct image resolution
- Reduce the file size.

Resize your images to correct image resolution

There are two methods to measure the size of an image. You can get the image height and width and calculate the physical image size as well as the number of pixels. Also, you can measure the file size by calculating the byte count.

Images that are not appropriately sized will cause serious rendering issues in mobile devices. The developers should make sure that they resize the images to fit into the available area before they deliver content to the mobile.

Since the display sizes vary from the smallest device to the biggest tablet, some developers save multiple size images and send the most suitable image to each device. On the other hand, using one image for all devices will give rise to some serious issues, especially in the small screens. When you double the width and height of an image, the image size increases by 4X, which causes delays in loading.

As mentioned earlier, the basic method that we can use to avoid this is to have an image for each and every screen size but, the developer has to create this manually. This is a superb solution but identifying each and every screen size is a hectic process. However, there are free tools available to profile every device and this will simplify this issue.

There is another method that developers use nowadays. They use CSS media quarries to manage images by manipulating the CSS code.

The following code shows how we can use a different image for the horizontal and vertical screens:

/* Portrait */

```
@media only screen
and (min-device-width: 320px)
and (max-device-width: 480px)
and (orientation: portrait) {
body{
background-image:url(images/bg-portrait.gif);
}

/* Landscape */
@media only screen
and (min-device-width: 320px)
and (max-device-width: 480px)
and (orientation: landscape) {
body{
background-image:url(images/bg-landscape.gif);
}
}
```

However, regardless of the challenge, the recommendation is to resize the images for mobile devices whenever possible. The properly resized image can save many bytes and will improve the user experience.

Reduce the file size

Reducing the overall data contained in an image file can reduce the image file size. This can be done by reducing the file size using a compression mechanism, or by physically reducing the file size by cropping it.

You can use two methods to compress an image file, lossy and lossless:

- Lossy compression can save up to 90 percent of the initial file size by removing information from the original file. It can give outstanding results with just a fraction of image quality lost.

- Lossless optimization keeps the original information intact, but it will push the image to the extreme to get the result. This option is good if you are concerned about the image quality, but this mode is time-consuming.

Image compression tools

We use various image formats for our websites. Whatever the format you use, it is essential to optimize those images properly.

Also, Google has developed a new image format called **WebP** that is supported in Chrome, Opera, and Android. The new format is optimized to enable faster and smaller images on the Web and it is about 30 percent smaller in size compared to JPG and PNG while the visual quality remains the same. Also, The WebP format has features that are present in other formats as well. It supports the following:

- **Lossless compression**: The lossless compression format is developed by the WebP team.

- **Lossy compression**: The lossy compression is based on the VP8 key frame encoding. VP8 is a video compression format created by On2 Technologies as a successor to the VP7 format.

- **Color profile**: It may have an embedded ICC profile.

- **Metadata**: It may have EXIF and XMP metadata (used by cameras).

- **Transparency**: The 8-bit alpha channel is useful for graphical images. The alpha channel can be used along with lossy RGB, a feature that's currently not available in any other format.

- **Animation**: It supports true-color animated images.

Because it results in better compression of images and has all these features, the WebP format could be an excellent replacement for PNG, JPG, and GIF.

Those who still use PNG, JPG, or GIF, can get the best out of it by optimizing those images properly. For image optimization, there are a number of free tools available on the Internet. You can download them; most of those tools are really easy to deal with.

Tiny PNG

According to the website, https://tinypng.com/:

> *"TinyPNG uses smart lossy compression techniques to reduce the file size of your PNG files. By selectively decreasing the number of colors in the image, fewer bytes are required to store the data. The effect is nearly invisible, but it makes a very large difference in file size!"*

Working with Tiny PNG is really easy. You just have to drag and drop your PNG or JPG file into their website, and you will get the optimized image.

To illustrate their optimization, I have uploaded five images, and I got the following result:

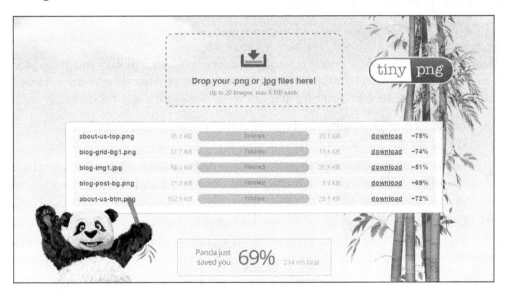

As you can see, I have just saved 234 KB only uploading five images to their website.

ImageOptim

There is another tool call ImageOptim, and you can download a free version from https://imageoptim.com/.

According to the website:

> *"ImageOptim is a free app that makes images take up less disk space and load faster, without sacrificing quality. It optimizes compression parameters, removes junk metadata and unnecessary color profiles."*

To illustrate its optimization, I have downloaded their tool and used the same five images that I have used in the previous example, and I got the following output:

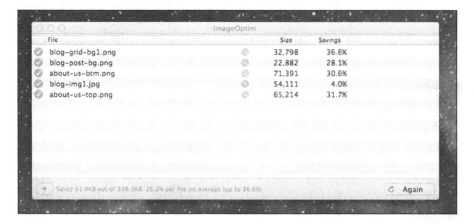

Kraken

According to the website `https://kraken.io/`:

> *"We optimize your images and accelerate your websites.*
>
> *Kraken is a robust, ultra-fast image optimizer and compressor with best-in-class algorithms. We'll save you bandwidth and storage space and will dramatically improve your website's load times."*

Kraken is another online tool that we can use to compress our images. It is a free service, and you can visit their website `https://kraken.io/`. Unlike TinyPNG, Kraken has a few options that we use to customize our image settings.

To illustrate their optimization, I have used the same five images, and their output is as follows:

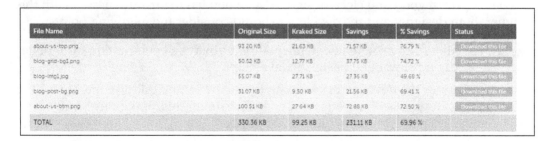

File Name	Original Size	Kraked Size	Savings	% Savings	Status
about-us-top.png	93.20 KB	21.63 KB	71.57 KB	76.79 %	Download this file
blog-grid-bg1.png	50.52 KB	12.77 KB	37.75 KB	74.72 %	Download this file
blog-img1.jpg	55.07 KB	27.71 KB	27.36 KB	49.68 %	Download this file
blog-post-bg.png	31.07 KB	9.50 KB	21.56 KB	69.41 %	Download this file
about-us-btm.png	100.51 KB	27.64 KB	72.88 KB	72.50 %	Download this file
TOTAL	330.36 KB	99.25 KB	231.11 KB	69.96 %	

Image optimization is really essential when we build a mobile website. There are a number of free tools available on the Internet that we can use to compress our images. Size does matter, make sure to use the correct image size and optimized images to build a mobile website. By doing so, you can give a better user experience to visitors.

Unnecessary contents

Mobile devices have had an exquisite journey so far, but there are still some issues with them, because of their physical limitations. There are different sizes of smartphones and tablets available in the market and to serve those devices, the developer should include flexibility into his design. However, many mobile users still find it a bit difficult to read mobile content because many developers try to add as much information on the mobile screen and most of the time they try to replicate the desktop screen in the mobile.

However, doing this does not only create performance issues for the mobile devices but it will also create some issues in user experience as well. So, when building a mobile website, developers should ask the following questions to themselves:

- What are my users going to do with my mobile website or application?
- What are the commonly used cases for my mobile application or website?
- What is the most important information that I should provide for my user?
- Am I delivering suitable content that is appropriate for my website, users, and mobile devices?

Clean design

Asking these questions upfront, developers can have a proper understanding of the content that they need, and they can present this to the designer so that the designer can come up with good and clean design. Also, when having the discussion with the designer, both designer and the developer should consider the following points:

- A smaller file size with adequate quality is always a better option for mobile devices and simplified design always helps.
- Sometimes the developers are going to use the same desktop content on mobile devices without any review. Before doing this, make sure your content is going to fit in your mobile device.
- Make sure that the font sizes are large and clean enough so that they are easy to read.

- Have a flexible design that would cater to all the device types including the smallest smartphones to the largest tablets.

- Consider all the scenarios that the user is going to face in your website, and then plan accordingly.

As I discussed earlier, many developers nowadays build responsive websites, and it is a trend. However, unlike mobile-only websites, a responsive web should have special attention, and it should be built carefully. To ease this process, the designers and the developer can use the following tips:

- Using a few images with smaller file size will help you to save screen space as well as it will reduce load time

- Use one column layout if possible and it will simplify the design

- Simple navigation will help you to use larger buttons and good for UX

- Always add a link to the desktop version, because some user prefers to do that

Duplicate content

When we browse a page, if we download an identical piece of content such as images, scripts, and stylesheets from the server, it will create duplicate contents. The duplicate content will slow down our website or application and will also create unnecessary load on the network as well.

To illustrate this properly, imagine a situation that you are going to send the same 5 KB image twice to your user. If your website has 4,000 daily users, you are sending an additional 20 MB data to your users. The battery consumption for these 4,000 additional downloads could be similar to draining 30-35 percent of phone battery.

To overcome this issue, we have to go through with our design and code properly, and we have to investigate our website or application thoroughly. Doing that we can identify the issues, which we have, and we can propose a better solution to prevent this and also, we can introduce a proper caching mechanism to the project.

We are going to discuss caching in the later part of this book. Until then remember that caching is really important because of the following reasons:

- Without any delay, you can access cached files immediately, which will give you a faster experience

- The cached file will save the battery of the user's device, and it will keep your users from leaving your website or app

- The cached file will save your users data because the users do not have to download files each and every time they visit your application

Because of the aforementioned reasons, it is essential to manage your website's content properly. Having simplified designs and a proper development plan will give you a chance to improve your content management, and it will help you to have a good caching mechanism.

Why design and UX are important

UX is not a new concept to the mobile world. Because of the limitations of the mobile devices many designers and developers give their attention to UX but, the question of compromising between performance and UX still seeks a clear answer.

The fundamental of mobile UX is very similar to desktop UX. However, the mobile technology is still evolving at a rapid pace, and the UX best practices are still emerging.

Mobile UX is all about how to keep the balance between, things that you are going to keep on the home page and how are you going to manage the rest of the contents. Also, it is really important that you provide a strong visual element to the user to help, how to navigate the site. As an example, you should clearly define the clickable area in your website or application. Without clearly defined buttons, the user will get confused, and they will spend more time trying to find the website's behavior.

The following are a few points that you can consider when you enhance UX in your application:

- Having a clear navigation will give a good user journey to the end-user. When building a mobile website, you should have clear vision whether you are you going to use a dropdown menu or traditional standard menu.
- Having a proper color scheme will enhance your site's UX. Especially, using borders and drop-shadows, the developer can define buttons, and more.
- Have a proper understanding about the animation, durations and how it's behaving in the device. Especially, if you have swipe or pulled menu or content, consider about the response time.
- Use proper input fields. With HTML5, you can define the input field type so, when you click on the input field, it will bring the numerical keypad or normal keypad.
- A Touch area should follow the fat finger rule and make sure not to encroach on surroundings.

A good UX will bring success to every website. Consider every scenario that the user is going to face in your site and don't leave any room where the user has to think about what they have to do next. It should flow like a river without any obstacles.

Summary

In this chapter, we have discussed the importance of reducing HTTP requests and how to do so by using techniques such as CSS sprites and combining files. Also, we talked about the enabling of Gzip on the server and its benefits. We have discussed the importance of image optimization and the tools that we can use to do so. We have also discussed a few free online tools that we can make use of, and found that they are easy to handle. After this, we discussed content management and why it is necessary to remove duplicate contents. Finally, we explored the importance of UX and its effects on a mobile site.

In the next chapter, we will take a look at the techniques that we can use to optimize our mobile website. The chapter will guide you on how to use HTML 5, CSS3, media quarries, and much more.

Summary

In this chapter, we have discussed the importance of reducing HTTP requests [illegible] how to do so by using [illegible] Sass [illegible] combining [illegible] we [illegible] the enabling of [illegible] and [illegible] We [illegible] the importance of image optimization [illegible] tools that [illegible] a few free [illegible] we can [illegible] they are easy to [illegible] faster [illegible] discussed [illegible] necessary to [illegible] Finally, [illegible] the importance of [illegible] and [illegible] site.

In the next chapter, we will take a look [illegible] improve our mobile [illegible] chapter will [illegible] on [illegible] HTML5 [illegible] media queries, and touch [illegible].

3
How to Optimize Your Mobile Website

In the previous chapter, you learned about the essentials of Mobile Web Optimization. I can assure you, the technics that you have learned so far are very powerful and very easy to implement, but the outcome they produce is massive. So, I would like to expand on that area more in this chapter.

When developing a website, the frontend developer plays a huge role. Having your development team apply frontend optimization techniques to your mobile website can dramatically improve the site's performance for mobile users. As I discussed in the previous chapter, many of those frontend principles are fairly straightforward to implement.

To go through with this chapter, I assume you have a basic knowledge of frontend web development. In this chapter, we are going to cover the following topics:

- Use of HTML5 and CSS3
- CSS animation versus JavaScript
- Iconic fonts
- How to use media queries
- Displaying none in CSS
- Video and images via media queries
- CSS preprocessors
- Minifying CSS and JavaScript

Use of HTML5 and CSS3

Performance and user experience can make or break your mobile app or website, so in today's consumer market it is important to focus on both. Nowadays in a mobile web environment, many developers struggle with choppy transitions, endless spinning, and periodic delays in tap and touch events. Developers are trying very hard to get closer to native behaviors, but to do that they have to use many hacks, resets, and third-party frameworks.

However, using HTML5 and CSS3 features, they can overcome this issue to some extent, and I would like to discuss some of these methods with you.

Hardware acceleration and the Graphics Processing Unit

The **Graphic Processing Unit (GPU)** is a specialized unit that was built to accelerate an image's output to a display. Generally, the GPU is very efficient and is more effective than the general-purpose CPU, and it can process algorithms efficiently. Most modern mobile devices now have advanced chipsets, and GPUs are an essential part of this.

In normal circumstances, GPUs handle advanced graphic calculations such as 3D modeling details and advanced diagrams. However, we can use a GPU to perform our primitive drawings such as DIVs, drop shadows, and backgrounds. Unfortunately, most frontend developers use third-party framework and scripts to perform those tasks without using CSS3 features, and the third-party frameworks use our device's hardware to perform those tasks.

As developers, we should make sure we avoid using a device's CPU and GPU as much as possible, and we should make the browser perform those actions for us. Preferably, the CPU will set up the initial animation and the GPU will only be responsible for compositing different layers during the process. In CSS3, translateZ, scale3d, and translate3d do this by animating elements in their own layer.

This is an essential step to follow when we travel down the mobile optimization path. Let me give you some reasons why we should use these features:

- **Memory allocation and calculation overhead**: This is a very critical point because some developers are only concerned with hardware acceleration and because of that, they create every element in the DOM. However, this will be an issue when you maintain your code.

- **Power consumption:** I have already discussed why the power consumption is important. It's a known fact that when we use hardware in a mobile, we consume battery power too. Usually, when we build an application or a mobile website, developers have to follow certain guidelines. So, it is essential to restrict a browser's access to a device's hardware as much as we can.

- **Conflicts**: Sometimes this feature may create unwanted behaviors in the application when we are trying to accelerate a section using hardware acceleration. So, we have to be very careful with this.

HTML5 form attributes and input types

HTML5 now has some new types of input fields and attributes, which are really effective. Most modern web browsers offer support for these new elements and using those, developers can reduce JavaScript and browser hacks.

This is a list of the new HTML5 input types:

- `<input type="email" />`
- `<input type="number"/>`
- `<input type="url"/>`
- `<input type="color" />`
- `<input type="time"/>`
- `<input type="datetime-local"/>`
- `<input type="datetime"/>`
- `<input type="date"/>`
- `<input type="month"/>`
- `<input type="range"/>`
- `<input type="search"/>`

This is a list of the new HTML5 attributes types:

- `autofocus`
- `autocomplete`
- `required`
- `pattern`
- `novalidate` and `formnovalidate`

The attributes mentioned here are very powerful, because earlier the developer had to use JavaScript or hacks to add a validator, color picker, date picker, and so on, but with these new fields and attributes, it's just one line of HTML code. However, some browsers still don't support some of the input types and attributes, and they will display this as a normal input field.

Using web storage in place of cookies

We have been using cookies to track users' data for years, but they have a serious disadvantage. The largest issue is that cookies data is added to each HTTP header request, and it creates a massive performance issue. To minimize that, we have to reduce the cookie's size, but with HTML5 we can use session storage and local storage to replace cookies.

Using CSS3 effects instead of requesting heavy images

CSS3 has many new styling options that we can use to replace images, and it will reduce the HTTP request as well. Also, if we can replace a 2 K image with 100 bytes of CSS, this is a massive saving.

I have listed some of the CSS 3 properties that we can use in the next few sections.

Border-radius for rounded corners

I still remember the early days when we got a button or a widget with rounded corners and we had to create images for each corner—it was so annoying. However, with the border-radius property, we can achieve this easily:

```
#element {
    border-radius: 25px 25px 25px 25px;
}
```

Box-shadow for drop shadows and glow

Using the box-shadow CSS property values for color, size, offset, and blur, developers can easily add multiple drop shadows on a box element (inner or outer):

```
.shadow {
  box-shadow: 0px 0px 35px 7px rgba(0,0,0,0.75);
}
```

Linear and radial gradients

CSS3 gradients allow developers to show smooth transitions between two or more defined colors. Back in the day, developers had to use multiple images for this requirement. But now, using the CSS3 gradients property, developers can eliminate images and minimize download time and bandwidth usage. Also, elements with gradients look better when zoomed in, because the browser generates the gradient.

In CSS3, developers can use two types of gradient:

- Radial gradients (defined by their center)
- Linear gradients (these go up/down/left/right/diagonally)

```
.background{
  background: linear-gradient(to bottom, #0071ea 0%,#ff0c0c 100%);
}
```

Transform properties for rotation

The `transform` property can be used to apply a 3D or 2D transformation to an element. This CSS3 property allows developers to rotate, move, scale, and skew elements:

```
.transform{
  transform: rotate(23deg) scale(0.937) skew(-9deg)
translate(3px);
}
```

Understanding CSS Filter Effects

The CSS3 filters are very powerful and developers can use them to show many visual effects. The CSS3 Filter property can be used to create effects such as color shifting or blur on an element's rendering before the element is displayed in the browser. The CSS3 filters are mainly used to adjust the rendering of an image, a background, or a border.

There are many CSS3 filters available at the moment:

* `blur()`
* `brightness()`
* `opacity()`
* `invert()`
* `drop-shadow()`
* `grayscale()`
* `hue-rotate()`
* `saturate()`
* `sepia()`
* `contrast()`

Using the following CSS code, you can use any of the available filters:

```
.filter {
  filter: <filter-function> [<filter-function>]* | none
}
```

An example of this is as follows:

```
.blur {
  filter: blur(10px);
}
```

CSS animation versus JavaScript

Having a nice animation or a text effect doesn't have a direct impact on your mobile site's conversion rate, but it will give your website an attractive look and feel. Nowadays, developers use CSS or JavaScript to create animations, and both of these have negatives and positives. Which method to use totally depends on the project and what kind of animation the developer is going to use. Anyway, I think CSS animations are excellent for simple animations such as toggling the UI element state, and JavaScript animations are good for complex effects such as bouncing, playing, stopping, and so on.

Most simple animations can be created using JavaScript or CSS, but the time you have to spend creating them will be different. So in my opinion:

- CSS animations are good for smaller, self-contained states for UI elements. For an example, when creating a navigation menu or a tooltip, developers can use a CSS transition property.
- If you need total control of an element, you should use a JavaScript animation. For an example, if you need a dynamic calculation or a complex animation, you should use JavaScript.

For an example, please look at the following image:

We are going to create two separate animations to convert this image using CSS3 and jQuery to the following:

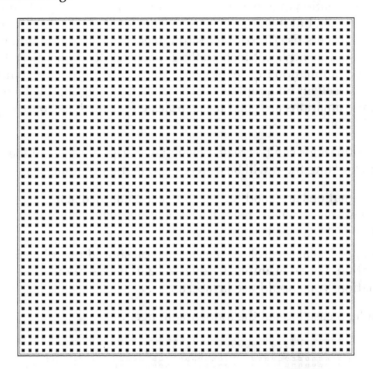

When we use CSS3 transitions, the animation is pretty smooth. The following image shows the full extent of the animation. There are a few of things that we can identify. Firstly, the frame rate has been capped, so unwanted repaints and calculations aren't done and only the area required is repainted, in this scenario, the rectangle surrounding the squares. The browser can choose the number of frames, how much data should change up front, and how to proceed. The animation is going to complete or stop/pause half way through and it would be difficult for us to start suddenly animating different properties half way through.

In this animation with CSS3, only around 40 events happened.

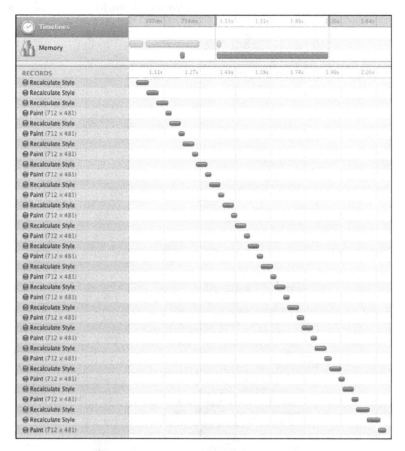

However, with jQuery animation when you try to do that, you will feel that the animation is not as smooth compared to the CSS3 version. This is mainly because the recalculate style is run for every element that needs to be animated and it has to recalculate around 9,500 styles during the animation.

As a result, only a small amount of repaints could be completed. The browser cannot foresee what's going to happen next because JavaScript could do anything at any time.

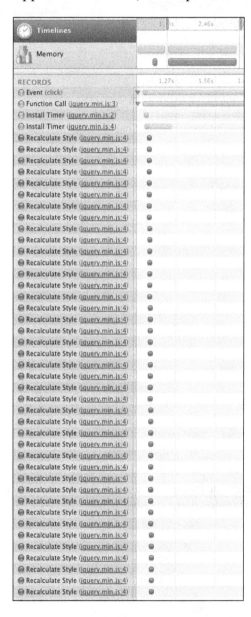

CSS animations

Without question, if you need to move an element from one place to another place, animation with CSS is the simplest solution.

The following is CSS code that will move an element 200 px on both the *X* and *Y*-axes. This is done by using a CSS transition that is set to take 1,000 ms. When the move class is added, the transform value is changed, and the transition begins:

```
.element {
  -webkit-transform: translate(0, 0);
  -webkit-transition: -webkit-transform 1000ms;
  transform: translate(0, 0);
  transition: transform 1000ms;
}

.element.move {
  -webkit-transform: translate(200px, 200px);
  transform: translate(200px, 200px);
}
```

Other than the transition duration, you can add options such as easing for the animation, so it will give you a smooth transition:

```
transition: transform 1000ms ease-out;
```

Also, you can combine this with JavaScript as well. All you have to do is create a separate CSS class and toggle that with JavaScript:

```
element.classList.add('move');
```

Having this kind of method gives your application a nice balance as you can manage the state with JavaScript and your browser will handle the animations.

Iconic fonts

In earlier days, frontend developers used raster files such as PNGs or JPGs when they needed to add an icon to a page. However, because of responsive web development, different mobile screen sizes, and retina display nowadays, it's hard to predict the screen size and as a result, the raster files appear pixelated. To prevent this issue, developers have to create multiple images to cater to different screen sizes, which has created a performance issue.

Luckily, using iconic fonts we can prevent many of the issues just mentioned, and it's much easier to use. Iconic fonts are just fonts that contain symbols and glyphs, and the developer can style them the same as regular text using CSS.

As well as this, using iconic fonts offers many benefits to us over raster files, and I have listed some of them here:

- Developers can easily apply any CSS property to iconic fonts, which gives total control to the developer
- Font icons are vectors, and because of that we can increase or decrease the size without losing the quality, which is especially effective for retina displays
- It reduces HTTP requests because we only have to make one or a few HTTP requests
- Iconic fonts are very small compared to images and as result, they load really fast
- Iconic fonts are supported in all mobile browsers

However, we can't use iconic fonts for all situations. As an example, if you are going to show a complex image with multiple colors, then an iconic font is not the solution. Iconic fonts are usually a single color and are designed according to a special grid.

In this section, I am going to discuss a couple of iconic font solutions in the industry, and they are really good.

Font Awesome icons

By the time I have written this book, Font Awesome will offer 585 icon collections either for free or for commercial or personal use. Developers can include Font Awesome in their project through CDN or manually by downloading it.

Once you download it, you have to add the `font` folder to your project and the `font-awesome.css` file to your CSS folder.

Then, call the `font-awesome.css` file in your project. Make sure you check that the `@font-face src` URL paths in your CSS file correlate to the suitable folder.

To use an icon, you should place it inside of an `i` element or a span. Then, add two classes to the element and you can call in any icon.

For example, you might want to call the camera icon as follows:

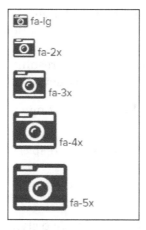

If so, you have to use the following code:

```
<i class="fa fa-camera-retro fa-lg"></i> fa-lg
<i class="fa fa-camera-retro fa-2x"></i> fa-2x
<i class="fa fa-camera-retro fa-3x"></i> fa-3x
<i class="fa fa-camera-retro fa-4x"></i> fa-4x
<i class="fa fa-camera-retro fa-5x"></i> fa-5x
```

IcoMoon icons

IcoMoon is another popular free iconic font solution in the industry. IcoMoon has a 2,000 plus vector icon collection and developers can search and download more than 4,000 plus icons. One of the options I like best about IcoMoon is that I can upload my own SVG image to my collection, edit it using their online app, and create a font.

Unlike Font Awesome, we can choose what fonts we need from their collection and generate them. Also, there is an option to download a SVG, PDF, or PNG as well.

Once we have downloaded IcoMoon, we have to follow the same method that we did in Font Awesome. Identical to Font Awesome, if you wish to use an icon, you have to place it inside of a span element.

How to use media queries

CSS media queries are an excellent way to deliver different content to different devices and screen sizes, giving the best user experience for each type of user. Media queries use media attributes to apply a CSS style to your website based on a device's properties, such as screen width, orientation, resolution, and more. So, when it comes to performance optimization, using media queries properly is essential.

A simple media query will look like the following:

```
<link href="css/mobile.css" rel="stylesheet" type="text/css"
media="only screen and (max-width: 768px)" >

<link href="css/tab.css" rel="stylesheet" type="text/css" media="only
screen and (min-width: 769px) and (max-width: 1024px) " >
```

As mentioned in the example, we are using two style sheets for our website: `mobile.css` for mobile devices and `tab.css` for tablet devices.

So, if your screen size is less than 768 px, the device will load and use `mobile.css`, and if the screen size is between 769 px and 1,024 px, the device will load and use the `tab.css` file. This way, developers can maintain different styles for mobiles and tablets.

Also, apart from `min-width` and `max-width`, there are some other properties that we can use as well:

Attribute	Output
`min-width`	Minimum width of the display area.
`max-width`	Maximum width of the display area.
`min-height`	Minimum height of display area.
`max-height`	The maximum height of display area.
`orientation=portrait`	Applied for any browser where the height is equal or greater than to the width.
`orientation=landscape`	Applied for any browser where the width is greater than the height. This is for any browser where the width is greater than the height.

I assume now you have an idea about CSS media queries. Let's see how we can use this in a real situation to enhance the performance on a mobile.

For example, when we build responsive websites, we often use a larger background image to cater websites. But, when it comes to mobile devices, we use a smaller background image because we can't show larger images on a mobile screen.

Normal CSS code will look like this:

```
/*Desktop*/
body{ background: url(background.jpg); }

/*mobile*/
Body.mobile{ background: url(background-mobile.jpg); }
```

When a developer writes CSS code, most of the time they call in images for both desktop and mobile, and whether we use them or not, the mobile device has to download both images, which affects performance.

However, using CSS media queries, we can avoid this very easily. All you have to do is define a breakpoint and call the background image relevant to the screen size:

```
/*Desktop*/
@media (min-width: 768px) {
  body {
    background: url(background.jpg);
  }
}

/*mobile*/
@media (max-width: 767px) {
  body {
    background: url(background-mobile.jpg);
  }
}
```

As you can see, we only added a few extra lines of code into our style sheets file. So now, devices that have a width up to 767 px (most mobile devices) will only load the `background-mobile.jpg` file, which is optimized for mobiles. Using this method, developers can call in optimized CSS properties for mobiles.

Displaying none in CSS

When we apply `display:none` rule using CSS we can hide HTML elements. Although you can hide an element from the frontend view using `display:none` property, this doesn't prevent the object from being downloaded to mobile devices. As a result, these elements will slow down your mobile site or application.

However, if you are going to hide an image from mobile devices and your intention is to remove it completely from mobiles, there is a method that you can use. For example, to hide an image from being displayed, we use the following code:

```
<div style="display:none;">
<img src="myimage.jpg" alt="" />
</div>
```

However, this method doesn't prevent it being downloaded to the device. To avoid this, we can use this image as a DIV background and hide it using CSS:

```
<style>
.imagehide {display: none;}
.mybackground {background: url(myimage.jpg) no-repeat; }
</style>

<div class="mybackground imagehide"> </div>
```

Using the preceding method, we can stop images from being downloaded to mobile devices without any issues.

Video and images via media queries

Media assets' weight is one of the biggest enemies of mobile devices. It says that roughly 60 to 70 percent of a site's weight is consumed by images, and this has been growing rapidly due to high-density displays.

When we build responsive websites, we remove the width and height attribute of an image, and we set the max-width 100 percent using CSS. By doing this, we make this image responsive and whatever the screen size is, the image will be resized for that screen.

Of course, this strategy requires developers to use images that are at least as large as the largest screen size at which they'll be displayed; if an image is expected as part of a layout, that could be anywhere from 320 px to 1,600 px, and the developer still requires to serve an image with an inherent width of at least 1,600 px. That's a tremendous amount of drained bandwidth and processing power for a mobile device, with no obvious advantage to the user. This bandwidth cost is increased by 400 percent when we update our assets to support HD displays. For an example, retina images are big in both dimensions, so this is four times larger than a typical image.

So what can happen is either it would be tremendously wasteful, or an older mobile browser might see all this data bearing down on it and fail completely, leaving the page unrendered.

However, using HTML5's video element, we can show assets that best suit different screen sizes:

```
<video>
<source src="video-large.webm" media="(min-width: 640px)"
type="video/webm">
<source src="video-large.ogg" media="(min-width: 640px)" type="video/
ogg">
<source src="video-large.mp4" media="(min-width: 640px)"
type="video/mp4">
<source src="video-small.webm" type="video/webm">
<source src="video-small.ogg" type="video/ogg">
<source src="video-small.mp4" type="video/mp4">

<!-- Fallback for browsers that don't support 'video': -->

<a href="video.mpg">Watch Video</a>
</video>
```

In the preceding example, the smaller of the two video files, in whichever format is supported by the browser, is displayed to any user with a display width less than 640 px. This property is well supported, and will work with current versions of Chrome, Firefox, Opera, Safari, Internet Explorer, iOS, Windows Phone, BlackBerry, and Android.

CSS preprocessors

If you are a frontend developer, you already know how important it is to write proper CSS for your website. Actually, CSS code doesn't have a direct relationship with performance optimization apart from style sheets' file size and duplicated classes. However, if you are planning to write code that is future-ready and easily maintainable, using a CSS preprocessor will ease your workload.

Of course, if you are working on a simple website, a preprocessor might not be required always. But, if you are working on a larger website and you have to deal with multiple style sheets and many CSS rules, a preprocessor will come in handy and will improve your code's quality.

In this section, I will briefly discuss SASS and LESS, the most popular CSS preprocessors available out there.

SASS and LESS

Both SASS and LESS are backward-compatible, and developers can easily convert their CSS files into LESS or SASS by just renaming the CSS file extension to `.less` or `.scss`, respectively.

LESS is based on JavaScript and SASS is Ruby-based, but when using these two, developers don't have to know anything about either language or use the command-line compiler to compile the files into CSS because there are many free applications available on the market to do the job. These applications will watch for any changes that you make to your `.less` or `.sass` file and they will automatically compile those changes and update your CSS.

There are many benefits for developers who use SASS or LESS in their project, and I will discuss some of them here.

Variables

When you deliver a completed project to a client, how many times did you get a request from a client that they need to change the main font to something else or they need to change the color scheme of the site? I often get this request from clients, and it's so frustrating because I have to go through each style sheet and make the change, and sometimes I miss some sections. However, by using SASS/LESS variables, we can eliminate this possibility completely.

SASS

Variables in SASS are defined using the $ symbol:

```
/* Variable for primary color*/ $primaryColor: #000000;
$base-font-size: 14px;

body {
     background: $primaryColor;
   font-size: $base-font-size;
}
```

LESS

Variables in LESS are defined using the @ symbol:

```
/*Variable for primary color* @primaryColor: #000000;
@base-font-size: 14px;

body {
     background: @primaryColor;
   font-size: @base-font-size;
}
```

As I showed in the preceding code, you can use the `$primaryColor`/`@primaryColor` value throughout the site, and once we change the value, it will apply to all places.

Partials

Partials allow us to modularize our CSS and keep things easier to maintain. In other words, developers can write different CSS files for different sections of a website and inject it into one `.sass` or `.less` file.

SASS and LESS

In the following example, we use two break points: one for mobiles and one for tablet devices. Using partials, we can maintain two separate files for each version:

```
@import "partials/custom/variables";
@import "partials/custom/settings";
@import "partials/custom/mixins";

@media #{$small-only} {
  @import "partials/mobile/checkout-mobile";
}
@media #{$medium-only-v} {
  @import "partials/tab/checkout-tab";
}
```

Mixins

Mixins are a very powerful feature that are offered by CSS preprocessors. They are similar to functions where we pass in variables as parameters so that we can give a dynamic feel to our website.

For example, if you need to create a rectangle, but use different sizes and different background colors, the variable values can be added to mixins, between the parentheses, as parameters. Then we add the variable name to the relevant property value. When separated by a comma, mixins can have one or more parameters.

SASS

We can use the following SASS code to create two different size rectangles:

```
@mixin rectangle($width,$height,$bgColor){
        width: $width;
        height: $height;
        background-color:$bgColor;
    }
    .box1 {
```

```
        @include rectangle(100px,50px, #ccc);
    }
    .box2 {
        @include rectangle(200px, 100px, #ddd);
    }
```

LESS

We can use the following LESS code to create two different size rectangles:

```
rectangle(@width,@height,@bgColor){
        width: @width;
        height: @height;
        background-color:@bgColor;
    }
    .box1 {
      .rectangle(100px,50px, #ccc);
    }
    .box2 {
      .rectangle(200px, 100px, #ddd);
    }
```

Here, I covered some of the features that SASS and LESS offers to us, but really this is just the tip of the iceberg. SASS and LESS are extremely powerful tools that a developer can use to do many extravagant things.

Minifying CSS and JavaScript

When we develop our websites, we often leave white space and comments in our code so humans can easily read and understand it. However, when the code file is being executed on our devices, excess white spaces and comments no longer convey a meaning to humans. So, if we can remove the extra characters from our files, it will reduce our file size and enhance the performance of our website. There are many tools available out there that we can use to remove unnecessary characters (minify them) from our code.

Minifying CSS

In the preceding section, I discussed CSS preprocessors. Both SASS and LESS have CSS minifying options that developers can easily use. The main advantage of CSS preprocessor minification is that developers are never going to touch the CSS file and only work with .sass or .less. So, they can keep the white space and comments in their raw file and when compiling, all the unnecessary characters will be removed. Apart from that, there are many online tools available and developers can paste their CSS code into that and generate a minified version.

Some online tools that developers can use are:

- `http://www.cleancss.com/css-minify/`
- `http://csscompressor.com/`
- `http://cssminifier.com/`

Minifying JavaScript

The same as CSS minification, there are many free online tools available to minify JavaScript, and some of the **Integrated Development Environments (IDE)** have a built-in option for this.

A few online tools that developers can use include:

- `http://www.cleancss.com/javascript-minify/`
- `http://javascript-minifier.com/`
- `http://jscompress.com/`

As I mentioned earlier, developers can use online tools or third-party applications to minify their code. By doing so, they can reduce the file size, allowing them to transmit and process quickly. As a result, developers can remove a few milliseconds from the app's loading time.

Summary

In this chapter, we discussed HTML5 and CSS3 and how to use their features to create performance optimization. In particular, we talked about the importance of hardware acceleration and GPU, and we used CSS3 effects instead of using heavy images. Also, we talked about CSS3 animations versus JavaScript animations and how to use iconic fonts instead of images. After that, we discussed how to use media queries and display none in CSS. Finally, we explored CSS preprocessors and the importance of minifying CSS and JavaScript.

In the next chapter, I will discuss caching and optimizing techniques that we can use to enhance performance much more.

4
Caching and Optimizing

In the previous chapter, we learned how to use frontend techniques to optimize our application or website. Actually, I believe all the frontend developers need to follow this guideline for all types of web development, and not just for mobiles. Some of the techniques that I have discussed in the previous chapter do not have a direct relationship with performance. However, this will create a well-structured environment, which will make it easier to focus on performance.

In this chapter, I will discuss a whole different area that is related to performance optimization, which is really essential and easy to understand. To go through this chapter, you need to have a basic understanding of HTML, CSS, and JavaScript.

In this chapter, we are going to cover the following topics:

- Caching
- File order of external style sheets and scripts
- Empty source and link attributes
- CSS and JavaScript frameworks
- How to optimize JavaScript
- Load only what is needed
- Reduce the number of DOM elements

Caching

I assume that you remember the three main pillars of a mobile device that we discussed in the *Three main pillars* section of *Chapter 1, Pillars of Mobile Web Performance Optimization*. Without a doubt, every developer should consider these three factors when they build a mobile website or application. Whether your site is for a small coffee shop or a larger online shop, you just cannot ignore these three factors, and this is why caching is important. It is an excellent way to download websites faster, saving some parts of them in your browser and then when you visit that site again you don't have to download or calculate that section again. Using the cache, the developer can improve speed, energy consumption, and user experience of his application or website.

Cache-Control

When we cache a file, it is available to reuse straightaway, which makes our application or website appear fast in performance. Also, enabling caching and using cache-control directives correctly, we can reduce unwanted data consumption and connections. These savings will help the user to stay under his data cap and will save his battery power from draining, and enhance his site's responsiveness of wireless network that have a limited bandwidth. Although it has many benefits, some applications or websites don't use the caching mechanism properly.

Through the cache-control general-header of each request and response message, the HTTP 1.1 protocol supports cache management. In the HTTP 1.1 protocol there are two caching mechanisms, that is validation and expiration.

Validation is the method that cross-checks cache data with the main server whether caching data is useable or not. When a server sends a full response to a client, the server attaches a validator to it and the client keeps that entry with the resource. When the client sends another full request to the server, it sends that validator with the request and then the server sends that validator with the new validator attached to new resources. If the two validators are identical, the server sends a nonmodified 304 message to the client, and the client uses the cached entry. This method saves a lot of bandwidth and time because, the client doesn't have to download resources from the server again. However, if the validator is not identical, the server sends full information so that client doesn't have to worry about seeing old content.

Using the expiration method, the client can completely avoid making a request to the server. This is done by the server setting up a specific expiration time for resources. Caching can check the expiration time and make the request accordingly. The server adds the expiration time using expiration headers or using max-age cache-control directives.

These two mechanisms determine how to get data from the server and how often it should be updated. However, to manage cache properly, a client or a server should provide explicit directives. We are using cache-control headers to do this task. In both caching methods, priority is given to the explicit directives rather than implicit directives in the caching-header. For example, max-age directives have the priority over the expiration time set in expires headers.

To implement a caching mechanism in your application, you can use a few methods. There are libraries available to enable this task, and some operating systems have an inbuilt function for this. However, the best way to do this is by incorporating a response entity cache functionality in your code.

Content prefetching

When delivering content to your user, you should always keep in your mind that wireless networks are slower than the wired networks, and this makes content delivery slower to the end user. Also, wireless users expect the same fast service as a wired connection. To balance these expectations, developers should have a proper content management plan, and that plan should have a choice, such as should we deliver the content to the user only if the user requested it? Or should we anticipate user behaviors and download the data before the user requests it?

Prefetching is the method of downloading and caching data before the user requests it. If we use this method wisely, we can speed up user experience on our website or mobile application.

In prefetching, we are going to use some predictions of contents that the user is going to request next. So, once we downloaded data for our initial request, the application begins to download data as we predicted, and it will store in our cache. By storing those prefetched data in the cache, the application makes sure to present those data quickly when the user made the request.

When designing the prefetched content, the developer needs to ask the following questions:

- For a different type of content, what are the goals?
- What workload will be used in testing?
- What will the underlying baseline system be for how prefetching is applied?
- Which key performance metrics are we going to use?

However, prefetching should be used wisely as I have mentioned earlier because of the following reasons:

- Sometimes the user's behavior is hard to predict
- The user may download content that he is never going to see, which creates unwanted overheads
- If the user has a data cap, unnecessary content will cause an issue to their data plan
- Analytics may become invalid because some content will be registered as being seen, but this is not the case
- Also, if you prefetched too much content into the user's device, it will consume the memory of their device, which may slow down its performance

Make favicon icon small and cacheable

I assume you already know what favicon is, and why we are using it on a website. Anyway, if you don't have an idea about it, the favicon is the small square image that is placed on left side of the browser's address bar, and it gives a graphical representation about your website. Usually, favicons are 16*16 pixel image:

We generally place favicons in our server root. We have to include a favicon in our website because the browser always requests this icon from the server, and if we don't send it then the browser will receive a 404 not found error. Because the favicons are on the same server, when the image is requested, the cookies will be sent each and every time. This image also plays a vital role in the download sequence. As an example, when we request extra components on page load, before the additional components get downloaded, the favicon will be downloaded to the client's mobile device.

So, when we use a favicon on our website, to minimize these issues we should use a small optimized image as a favicon. Also, we can set expires headers as we have discussed in the previous section as well.

File order of external style sheets and scripts

When we render a page, the order of the CSS and JavaScript's files that are located in the code have a direct impact on how quickly that we can show the page to the user. If we can load the CSS files before the JavaScript files, the page can start the rendering straight away, and we can download other files parallel to the rendering which increase the website rendering speed.

On the other hand, if we wait to download CSS files after the JavaScript files, the website will have to wait until all the JavaScript files get downloaded to start the page rendering. Another issue with this is that if a JavaScript code has a CSS code dependency, there will be a conflict.

To minimize the impact, the recommended method is to load the style sheet files first, then push the scripts to the bottom of the page as much as possible.

Empty source and link attributes

If we have an HTML tag containing an attribute, without value, we say it is an empty attribute. The issue with this empty attribute is, if the empty attribute is a source or a link, some browsers will still try to connect to the server, even we set it as an empty value. This unwanted request and overhead will create delays in our mobile website or application. As an example, take a look at the following anchor tag and image tag:

```
<a href="">
<img src="">
```

Also, take a look at the following JavaScript code:

```
var image = new Image();
image.src = ""
```

Usually, the HTML recommendation is that if there is an src attribute, it should contain a URL. However, if you are using HTML5, you can have an empty attribute because, HTML5 uses specified algorithm to avoid extra requests if there is an empty attribute.

But, if you are not using HTML5 different browsers will behave differently:

- Some browsers make the request to the same page
- Some browsers ignore the request if it's an empty tag
- Some browsers make a request to the file location directory

As a result of these different behaviors, it will create an unwanted overhead in your website or application. The most critical section that we have to address is that sometimes content management systems and template engineers forget to add URL to the attribute and without the developer's knowledge the website or application will have empty attributes.

So, as a practice you need to always be careful with the following tags:

- `href src`
- `img src`
- `iframe src`
- `script src`
- `link href`

CSS and JavaScript frameworks

If you are building a responsive website, for mobiles as well as for desktop, I think you need to use a CSS/JavaScript framework. Some people can argue that using a framework can decrease your site's performance because of the unwanted component that bundles up with the framework. There are some components in the framework that you are never going to use. However, I have a different opinion in this regard. Most of the frameworks now have the option to choose. With the components that we are going to have and as per our requirement, we can customize our framework. Using this customizing option, we can get the following benefits of frameworks:

- You don't have to be a professional programmer to build a website
- Using a framework will drastically cut down our development time, and you can invest that saving on performance optimization
- You can plug in new components with ease
- The framework code is well tested and has a lot of community support
- You will get regular updates and new features for the framework
- As I have mentioned earlier, because of the community support and regular updates, you don't have to worry about the security

There are many CSS and JavaScript frameworks out there to choose from (`http://www.awwwards.com/what-are-frameworks-22-best-responsive-css-frameworks-for-web-design.html`), and I will try to list down a few of them here.

Bootstrap

The framework was originally created by a designer and a developer at Twitter in mid-2010. Since then, Bootstrap has become one of the most popular, successful frontend frameworks and open source projects in the world. If you are familiar with LESS, you can work with Bootstrap easily because it has a LESS version, which is easy to customize.

If a developer needs to customize anything within the framework, it can be done easily because the framework is built on 12 responsive column grids, layouts, and components. For both fluid and responsive width layout, nesting and offsetting are possible. Also, using responsive utility classes the developer can make a certain block of content visible or hidden only on devices based on their screen size. This proves to be very useful when a developer prefers to hide some content based on screen size.

Adding a class such as `.visible` to an element will make it visible only to mobile users. There are similar classes for tablets and desktops.

Zurb Foundation

Zurb Foundation has its own styleguide with a collection of HTML and CSS. It helps developers to build amazing websites very quickly. The framework now is very powerful and popular, and millions of developers use it to build.

The Foundation framework has many benefits. The framework is built with compass; it allows the abstraction of presentation to be easier and faster. Also, it has no extraneous IDs, classes, or non-semantic empty HTML elements. The framework contains cleaner markups, so it's much easier to maintain and reuse. Also, Foundation is built with the mobile first approach, in the true spirit of progressive enhancement.

Foundation is optimized for mobile devices. The framework's newest version takes it even further with greater support for touch input. Hardware acceleration has added to the framework for smoother transitions and animations. It makes your application more native on your mobile.

UIkit

UIkit was developed by YOOtheme, and the current version is 2.21.0. Compared with other frameworks, UIkit is lightweight and the framework was created with an eye on modern trends. There have been responsive design and grid sets - the underlying trend, so they have been released in UIkit as its core features. Such technologies aren't surprising nowadays, but these features are made on a really high level.

Semantic-UI

This is another new framework in the market, and its popularity is increasing day by day. This framework gives a very good competition to Bootstrap and Foundation with its new advanced features. It comes with many integrated third-party libraries such as Ember, Angular, and Meteor. With these libraries, you can build a large and complex website with ease.

Susy

Susy has a really advanced grid system. Unlike other frameworks, Susy doesn't have a prebuilt grid system. However, using Susy mixins, you can make any kind of grid, and you don't have any restrictions. Also, if we compare Susy with Foundation or Bootstrap, Susy doesn't have a very large package, but Susy is all about the grid.

jQuery

Without a doubt, jQuery is the most popular JavaScript library for developers. jQuery was released in 2006 as an open source project and since then, its popularity is growing at a rapid rate. Working with jQuery is really easy if you are familiar with HTML, CSS, and JavaScript.

Also, we can now use the jQuery mobile framework to create a mobile application as well. jQuery mobile is a user interface framework built on top of the jQuery library. The most important part of the jQuery mobile is that it's used only HTML5 standard code.

These are the features of jQuery mobile:

- Open source and free to use
- Cross-browser, cross-platform, and cross-device compatible
- The user interface is optimized for touch devices
- Only uses semantic HTML5 code
- AJAX calls are made automatically to load dynamic content
- Lightweight in size
- Accessibility support

AngularJS

AngularJS is a JavaScript framework developed by Google to simplify the frontend development. If you are building single page apps, AngularJS would be the best option for you. When you write AngularJS code, you can modularize it, which means you can follow MVS patterns, and it will increase reusability of your code. Also, it's much easier to test and maintain.

Another option that AngularJS presents is that we can declaratively bind our model to HTML elements. So, when the model is changed, the view will change automatically and vice versa. This can reduce a lot of unwanted code lines.

Also, the best thing about AngularJS is that many developers can work simultaneously without an issue. Adding to this, AngularJS offers features such as routing, filters, and animations as well.

Ember

Ember.js is another open source JavaScript framework built and based on MVS patterns. It allows developers to create a single-page web application with ease. Currently, Ember has 400,000 downloads per month, and many modern applications use Ember as their framework. Ember always uses future technology in today's applications, keeping the backward compatibility. Similar to Angular, Ember has its own routing system represented by a URL. Also, Ember templates are used to create the application's HTML.

Aurelia

Aurelia is another JavaScript framework in the market, and it's an official product of Durandal.Inc. Aurelia is a completely free and open source framework, and it can be customized easily. It has no external dependencies except polyfills, and it uses future technology to build today's applications. One of the main features of Aurelia is that it has two-way data binding to any object. Also, it can be used to extend HTML elements and developers can create custom HTML elements with ease.

Aurelia has an advanced client-side router with its pluggable pipeline, dynamic route pattern, and asynchronous screen activation. Also, applications built using Aurelia can be tested quickly.

Knockout.js

Knockout.js is another popular MVC based JavaScript framework, and it has a convenient data binding feature similar to Angular JS. Knockout.js is best for prototyping small applications or it can be used to introduce data binding to a legacy code. It's developed and maintained as an open source framework by Steve Sanderson, a Microsoft employee, and as he said, *"It will stay as it is in future as well."* The frameworks have declarative data binding, automatic UI refresh, dependency tracking, and templating systems.

How to optimize JavaScript

Nowadays, we use JavaScript in our application or website to create visual effects. Sometimes, it creates these visual effects by manipulating styles or doing calculations such as sorting or searching. However, if we don't use these properly, JavaScript can cause performance issues on our website or application. To minimize these impacts on the website, you should consider the following:

- Avoid using `setTimeout` or `setInterval` for visual updates. Instead of these you can use `requestAnimationFrame`.
- Move long-running JavaScript main thread to Web workers.
- Use browser development tools to profile JavaScript and assess the impact.
- Use micro-tasks to make DOM changes rather than using several frames.
- Keep your HTML code clean by removing unwanted `DIV` and `SPAN` tags.
- When updating styles, do them as a batch.
- Build DOM separately, before it is added to the page.
- Remember to unbind events when they are no longer needed.
- Learn about event bubbling. Use `jQuery.bind()` instead of `jQuery.live()` and `jQuery.delegate()`.
- Avoid creating unnecessary functions.
- Learn and use native JavaScript functions and constructs.

Load only what is needed

With the help of AJAX we now use lazy loaders in our websites. The advantage of lazy loading is that we can load the resources when we need them rather than having them around all the time. Using this technique, we can boost up the website's loading time and stay under the user's data cap. For example, we don't have to load large images in mobiles that are intended for the desktop version, and we don't have to load scripts on Android devices if it's only meant for iOS.

JavaScript is good at testing for support and then loading resources on demand. However, we don't use it properly. At the moment, we use more and more solutions that load large amounts of high-end resources because we assume that caching will improve the experience as the user moves through the site. Actually, this is a waste of time as it doesn't help the users who will never benefit from that high-end experience. It might not seem to be a problem for people who have a fast connection. However, we are not building applications or websites only for those people. So, the more we can delay loading unnecessary content or subsequently storing it on the user's device, instead of repeatedly loading it, the better our solutions will be.

I think we need to avoid using preloading content as much as possible. To do so, consider before adding 10 different fonts for your homepage, the CSS framework that we use only to create a two-column layout, a JavaScript library only to create a button event handler. Instead of doing all these, check the available screen space before loading content and hide the elements if these are located on the outside of the screen. So, the user actually scrolls them into view we load them in the screen.

There is natural downtime in the interaction with our apps. For example, people will spend some time entering data into forms. So, why not use that time to load additional resources? A focus handler on the first text field could trigger good resources download. If the user never enters the form, nothing needs to happen.

Reduce the number of DOM elements

If we have a sophisticated website, that means we have to download more bytes and slower DOM access in JavaScript. There is a difference between looping through 500 DOM elements versus 5,000 DOM elements, if you need to add an event handler to an element.

If a page has a huge number of DOM elements, it means that there is room for improvement. So, you should go through with your code and enhance it without removing valuable content.

Using a CSS framework such as Bootstrap or Foundation can eliminate those issues, because they have a proper structure. Also, using the `reset.css` style sheet file, the developer can remove default browser formatting. Additionally, if you are going to use a new `HTML` tag, use it only when it makes sense semantically, and not because it renders.

Finally, testing the number of DOM elements on a page is easy. Simply type the following command in browser toolbar console:

```
document.getElementsByTagName('*').length
```

Summary

In this chapter, we have discussed the importance of cache and how the caching mechanism works. Having a proper caching plan can boost up your site performance immensely. After this, I have explained how a developer should call JavaScript and CSS files and why we should avoid empty source and link attributes. Then, I have gave a brief introduction to CSS and JavaScript frameworks. These frameworks are really powerful, and we can use their advanced technology for our benefit. In the later part of the chapter, I have explained how we can optimize JavaScript to gain performance and why we should load what is needed. Finally, I have discussed the importance of reducing DOM elements.

In the next chapter, I will discuss how to monitor and debug issues in an application or mobile website.

5
Monitoring and Debugging Our Website

In the previous chapters, we went through some of the techniques that we can use to improve our website's performance. Actually, most of the methods that we discussed were straightforward and really easy to implement. As an example, optimizing images, minifying CSS and JS, and using CSS3 and HTML as much as possible is really doable with minimum front-end development knowledge. However, using caching techniques, optimizing JavaScript, and choosing a CSS framework for your website requires some expertise in the domain. So, if you are going to build a website by hiring a third-party agency or a freelancer, you should always consider these factors. Because, every developer doesn't have the skills to build for performance.

In this chapter, I am going to discuss an interesting area. That is, how to monitor and debug your mobile website. Actually, you don't have to be a developer to go through with this chapter and use the methods that I am going to show, as you can test your own website and evaluate it by yourself. However, to go through with this chapter you should at least have some knowledge of web browsers and how they work. Also, the tools discussed in this chapter require hands-on practice. So, I encourage you to practice with those tools as much as possible.

In this chapter, we are going to cover the following topics:

- Profiling tools
- A browser's DevTools performance
- Performance tools in Firefox, Safari, and IE
- A Google Chrome emulator
- Google PageSpeed Insights
- YSlow

Portions of this chapter are modifications based on work created and shared by the Android Open Source Project and is used according to terms described in the Creative Commons 2.5 Attribution License.

Profiling tools

Normally, mobile devices use four primary pieces of hardware to render a web page on to your screen. The CPU calculates all the display units, the GPU renders all the images to the screen, the memory stores all the data, and the battery provides the necessary electric power. All of these hardware units have their own limitations. Forcing or exceeding those limitations will cause your application or website to become slow, have a rendering issue, or will drain the battery.

To discover the reasons for specific performance issues and to find a solution, you have to take a look at your application's backend layer, use a tool to collect data about your application's behavior, generate a report with graphics, study those reports, and enhance your code.

Nowadays, many mobile development platforms and mobile devices provide profiling tools to generate visualized reports such as rendering, computing, and the battery performance of your application.

GPU Overdraw Walkthrough

Using a color coding interface, GPU Overdraw Walkthrough shows elements based on how often they are drawn underneath on your mobile devices. This tool helps us to identify where our application is doing heavy rendering work and helps us see where we can improve rendering overhead. To use this profiling tool, you should enable developer options on your mobile device.

To enable GPU Overdraw Walkthrough on your mobile device:

1. Go to **Settings** on your mobile device and tap **Developer Options**.
2. Select **Debug GPU Overdraw** in the **Hardware accelerated rendering** section.

3. Select **Show overdraw areas** in the **Debug GPU Overdraw** popup.

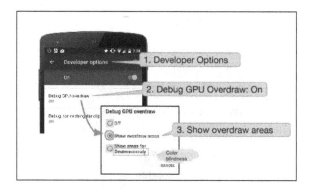

4. You will see that your device has changed to a delirium of colors. The colors will help you to diagnose your application's display behavior.

5. The colors indicates the overdraw area on your screen for each pixel, as follows:
   - ° **True Color**: No overdraw
   - ° **Blue**: Overdrawn once
   - ° **Green**: Overdrawn twice
   - ° **Pink**: Overdrawn three times
   - ° **Red**: Overdrawn four or more times

6. Anyway, some of the overdraw area is unavoidable. However, you should aim to arrive at a visualization that shows true colors and a 1X overdraw in blue mostly.

GPU Rendering Walkthrough

Profile GPU Rendering Walkthrough gives you a graphical representation of how much time it takes to render the frames of a UI window against the 16 ms per frame benchmark. This tool is good if you want to observe how a UI window performs against a 16 ms per frame target, and using the processing time, we can identify any exceeded section of the rendering pipeline. Also, using this tool, we can monitor spikes in the frame rendering time associated with a user or a program action. To use this tool, you should have at least Android 4.1 and have enabled the developer options on your device.

1. Navigate to **Settings | Developer Options** in your device.
2. Select **Profile GPU Rendering** in the monitoring section.
3. In the **Profile GPU Rendering** popup, select **On screen as bars** to overlay the graphs on the screen of your mobile device.
4. Go to the application that you want to profile.

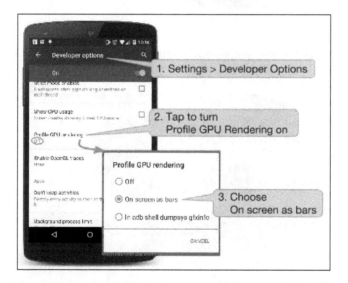

Once you activate this profiler, you will see a separate graph for each visible application:

- The vertical axis shows the time per frame in milliseconds and the horizontal axis shows the time elapsing.

- When you interact with your application, vertical bars will appear on the screen from left to right, graphing the frame's performance over time.

- Each vertical bar represents one frame of rendering. The taller the bar, the longer it took to render.

- The green line marks the 16-millisecond target. Every time a frame crosses the green line, your app is missing a frame, and your users may see this as stuttering images.

In the profile GPU rendering graph, each color in the colored sections of the graph represents each rendering pipeline.

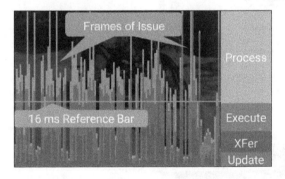

Here is the brief description about the preceding image:

- The green line indicates the 16 milliseconds mark. To achieve 60 frames per second, the vertical bar should remain below the green line, as any frame above the green line may cause pauses in the animations.

- The blue section of the bar indicates the time taken to create and update the view's display list. If the section's bar is tall, there may be a lot of custom view drawing or the onDraw method may have a lot of work to do.

- The purple section of the bar represents the time spent transferring resources to the render thread.

- The red section of the bar represents the time spent by Android's 2D renderer issuing commands to OpenGL to draw and redraw display lists. The height of this bar is directly proportional to the sum of the time it takes each display list to execute — more display lists equals a taller red bar.

- The orange section of the bar represents the time the CPU is waiting for the GPU to finish its work. If this bar gets tall, it means the app is doing too much work on the GPU.

Anyway, although this tool is named Profile GPU Rendering, all the processes mentioned actually happen in the CPU.

A browser's DevTools performance

In the Google Chrome web browser, Developer Tools (press *F12* or *Ctrl + Shift + I* to open the DevTool) provides an overview report about your web application's loading time, such as how long the browser has taken to process DOM events, paint elements to the screen, or render the page layout. It allows you to go deep into your application's events, frames, and actual memory usage, and it will help you to identify the root causes of your application's slowness.

Right now, we are going to have a look at the frame mode, which allows you to see how the browser performs when generating a single frame. By default, the timeline won't show any data, and you can record a session with the tool by opening your application and clicking on the record button (the grey circle) on the screen. You can use the *Command/Ctrl + E* shortcut as well.

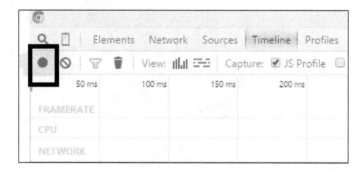

When you click the record button, it will turn to red and will start to capture the timeline of your page. Complete a few actions inside your application (button click/scrolling, and so on) and after a few seconds, it stops the recording.

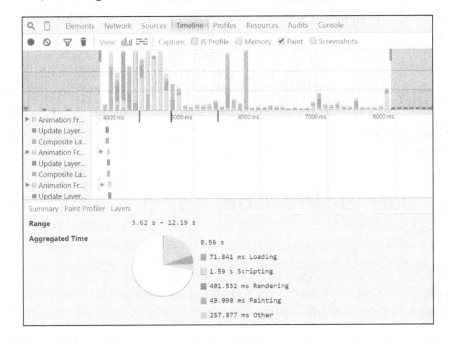

Clicking on a record will display extended information about the record, and using that information you can improve the particular event.

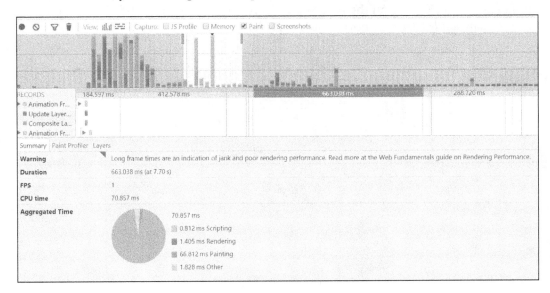

At the top of the timeline, the summary view displays horizontal bars representing the HTML and network parsing (blue color), JavaScripts (yellow color), painting and compositing events (green color), and style recalculation and layout (purple color) for your page. When we make visual changes such as scrolling or resizing, it will ask the browser to repaint the browser events. If we change the CSS properties, then recalculation occurs, while layout events (or reflows) are due to changes in the element's position.

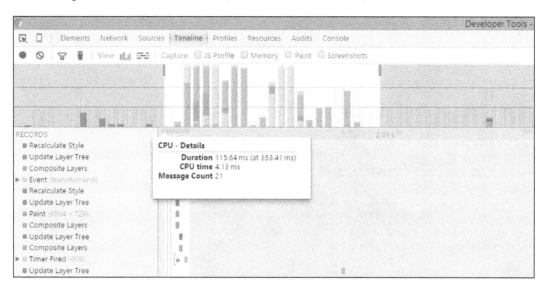

In Chrome, there are a few other tools available for us to use, so before going into tweaking rendering performance issues, I would like to discuss those as well.

In Google Chrome web browser, DevTool has a shortcut allowing developer to toggle the visible:hidden of an element. When visibility: hidden is applied to an element, it is not going to restructure the whole DOM tree, because the element will available in the page layout as an unchanged element. To use the shortcut, open the Developer Toolbar (*F12*) and select a DOM element using the element panel that you wish to hide and press the *H* key.

When coupling this with the paint rectangle and timeline, we can easily observe which item takes a longer period to paint.

Some of the user's interactions cause style changes to the DOM elements and changes to the DOM nodes. As a result, sometimes the browser has to repaint some of the areas. To understand why repaints occur, we can use the **Enable continuous page repainting** feature by navigating to **Drawer | Rendering**.

This feature in the **Drawer** panel helps us to identify elements that have a high paint cost on the page. It forces the page to do continuous repainting, providing a counter showing how much paint work is being done. You can use the *H* shortcut I mentioned earlier to toggle different style to observe what is causing the issue.

Another feature that we can use in the Chrome Developer Toolbar is the **Show composited layer borders** feature. You can activate this in the same location, and this feature allows us to see the DOM element that is being manipulated at the GPU level.

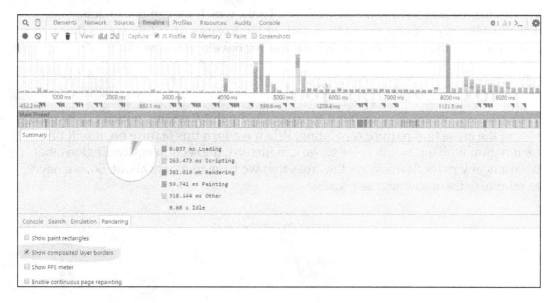

If an element gets the advantage of GPU acceleration, there will be an orange border around the element.

Remote debugging

It's a bit hard to build a mobile website or an application without debugging it on actual devices. Luckily, nowadays modern web browsers provide the necessary tools to debug your websites on the desktop as well as for mobiles. Using these tools, developers can connect their device to a PC and run a debug session.

Normally, we can remotely debug by connecting our mobile device to our PC using a USB cable. When you connect your device to the debugging PC, you can profile pages using the timeline, and you can view and edit HTML, CSS, and scripts until you have the optimized result.

As I mentioned earlier, every major browser such as Chrome, Safari, Opera, and Firefox now has a remote debugging facility, and setting it up for your mobile device can be done in a few steps. So, please learn how to connect your device for remote debugging, which you can learn about by reading the remote debugging guidelines.

Performance tools in Firefox, Safari, and IE

We discussed Google Chrome's Developer Tools, which we can use to debug our website. However, if you are an IE, Safari, or Firefox user, don't worry, as those browsers offer their own developer's toolbar for developers as well.

Firefox Developer Tools

The same as Chrome, Firefox has a feature called **paint flashing**, which can be used to find the areas that require repainting. When we turn this feature on, it will tint each region with a random color so we can identify the section easily. Regions that have a heavy paint flashing are the areas that we have to worry about. So, we have to minimize them as much as possible.

To enable paint flashing:

1. Open Firefox.
2. Type `about:config` in the address bar and press *Enter*.
3. Accept the warning.
4. Right click and select **New | Boolean**.
5. Type `nglayout.debug.paint_flashing`.
6. Set the option to **True**.

You will see something like the following screenshot when you turned on `paint_flashing`:

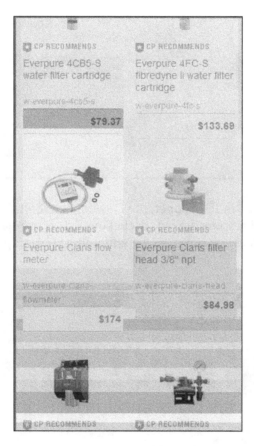

IE 11 Developer Tools

Surprisingly, the IE 11 Developer Toolbar has many new features including special UI responsiveness to profile performance, slowness, and CPU and memory usage.

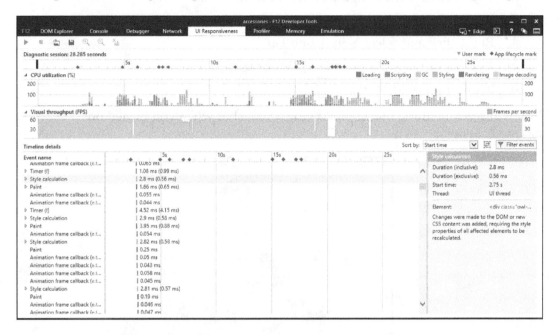

When you load the page in IE 11, open the Developers Toolbar (*F12*), and start a new profiling session on UI responsive tab by clicking the record button. Once clicked, don't do any fancy actions on the page, so you will be able to pinpoint the exact reason for the issue.

Once you have completed the required action, stop the recording and the toolbar will generate a report for you. Using the report, you can find out which area has issues and what actions have caused them.

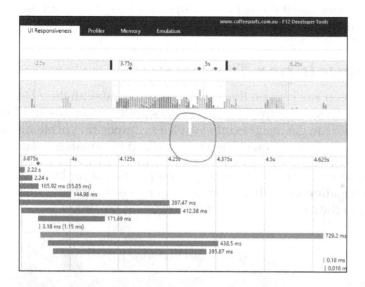

Also, IE 11's Developer Toolbar has another feature called **Network**. Using this, you can find out which requests you made when you opened a page, and how much data and time are consumed to complete each action.

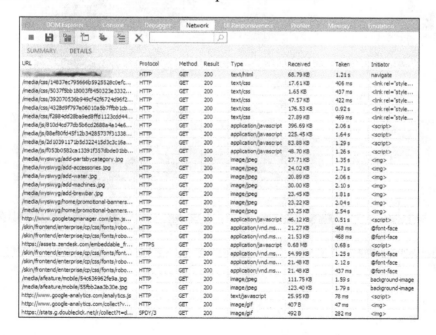

Not only these, but the IE 11 Developer Toolbar also has many other features available, and using these features, we can easily observe, HTML, and CSS phrasing, network requests, image decodes, script evaluation, animation frame callbacks, and more. To learn more about the IE11 Developer Toolbar, check out the official IE 11 Developer Toolbar guidelines.

Safari Developer Toolbar

Apple has also been working on their Safari browser and with their Developer Toolbar, we can track and improve our website's performance easily. Using the layer details sidebar, we can get insights into WebKit's compositing of elements. When you select a layer in the sidebar, you can get summarized information related to that layer.

The same as Chrome, you can also display compositing layer borders, generated in DOM tree's navigation bar, which overlays your page to provide a clear visualization of the layers and the number of times they are being repainted.

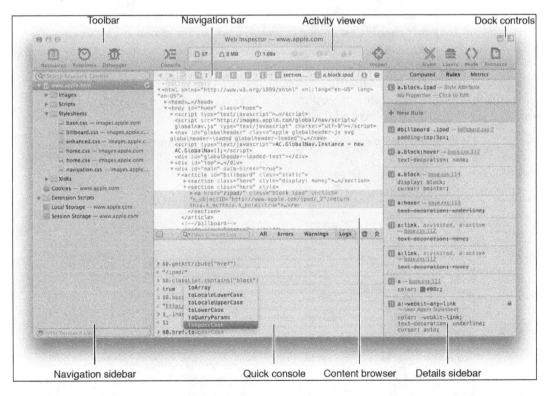

The following are the reasons for compositing:

1. Open the website in Safari
2. Open the web inspector
3. Click on the layers button
4. When you hover over the element, you can see the reason for the layer's promotion

You can also show the number of times a layer was repainted. This is helpful so you can understand which part of your page is getting excessively painted because of your scripts.

The Google Chrome emulator

We have already discussed a few tools that we can use in Chrome's Developer Toolbar to test a website's performance. Now I am going to discuss how we can use the device mode in Developer Toolbar. The benefits of this feature are that we can use our browser's viewport as a device emulator, and we can test our website's responsiveness.

To turn on the device mode, open Developer Toolbar and click on the toggle device mode icon. When the mode is activated, the icon turns blue and the viewport will transform into an emulator.

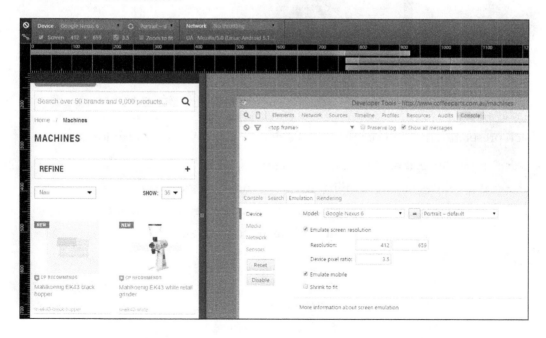

You can use Chrome's DevTools device mode to:

- Check the responsive design on different screen sizes and resolutions including retina displays

- Inspect, edit, and visualize CSS media queries

- Simulate the device input for orientation, touch, and geolocation

- Enhance your current debugging workflow by combining the existing DevTools with the device mode

As I mentioned earlier, using the preset, you can easily switch to any device that is in the list.

Each preset automatically configures the device's emulation in the following ways:

- If available, it enables touch emulation
- It emulates mobile scrollbar overlays and meta viewport
- It auto sizes (boosts) text for pages without a defined viewport
- It specifies the **User Agent (UA)** string for requests
- It sets the device resolution and pixel ratio

Also, using the network dropdown options, you can choose various network speeds such as 3G, Edge, DSL, offline, and more. This is really beneficial if you want to observe how your website responds and unfolds under different Internet speeds.

Google PageSpeed Insights

Google PageSpeed Insights (`https://developers.google.com/speed/pagespeed/insights/`) can be used to measure the performance of a page for mobile devices. PageSpeed fetches the URL twice, once for the desktop user agent and once for the mobile user-agent.

The PageSpeed gives a score of 0 to 100 points to your page, and a higher score is better. If you can get more than 85 points on a page, it will indicate that the page is performing well. However, this tool is being continually improved, and the score will differ from time to time.

PageSpeed Insight gives feedback on two areas, how to improve the page performance:

- **Time to above-the-fold load**: How much time has been consumed from when a user made the request and when the browser rendered the above-the-fold content (portion of a webpage users see on their screen before they scroll)

- **Time to full page load**: The total consumed time from the first user request to when the completed page is rendered by the browser

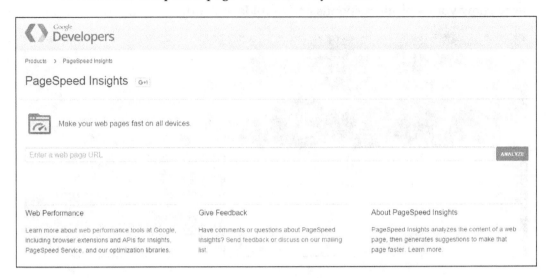

Once you run your page's URL using the tool, you will get the page's score and some of the recommendations that you could implement for your page.

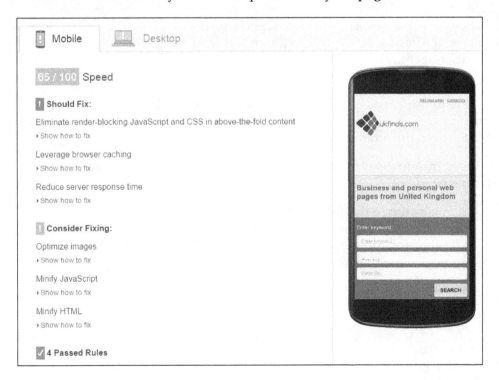

However, you have to consider that the performance of the network connection that you use has a direct impact on the page's score.

YSlow

YSlow is a tool developed by Yahoo! and it examines all the components of your page's performance and offers suggestions for improvements. YSlow is integrated into the Firebug web development tool for Firefox, and for Chrome you can download their extension.

The Yahoo! team has identified 34 rules that affect a web page's performance and YSlow is based on 23 of these 34 rules. Studies have shown that if a developer can improve the following 23 rules, they can enhance web performance from up to 25 to 50 percent.

- Minimize HTTP requests
- Use a CDN
- Avoid an empty `src` or `href`
- Add expires or a Cache-Control Header
- Gzip components
- Put style sheets at the top
- Put scripts at the bottom
- Avoid CSS expressions
- Make JavaScript and CSS external
- Reduce DNS lookups
- Minify JavaScript and CSS
- Avoid redirects
- Remove duplicate scripts
- Configure ETags
- Make AJAX cacheable
- Use `GET` for AJAX requests
- Reduce the number of DOM elements
- Ensure no 404s
- Reduce cookie size
- Use cookie-free domains for components

- Avoid filters
- Do not scale images in HTML
- Make any favicon.ico icons small and cacheable

When analyzing the YSlow page test, it gives a certain number of marks for each point, and based on the total marks, it gives an overall grade to your page.

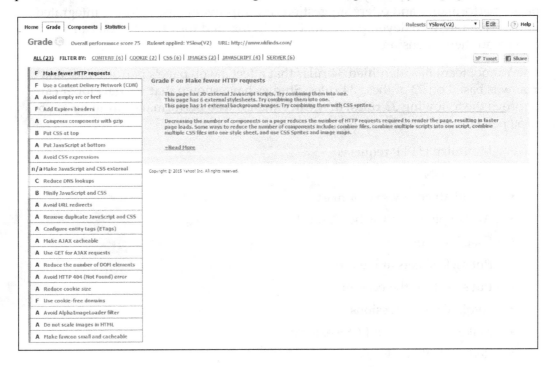

Summary

In this chapter, we discussed how to use profiling tools such as GPU Overdraw Walkthrough and GPU Rendering Walkthrough to debug and enhance our website's performance. After that, I explained the features of browser's DevTools and how we can remotely debug our website using actual devices connecting to our PC. Also, I explained the Firefox, Safari, and IE developer toolbars and how we can use those for debugging.

In the latter part of the chapter, we went through the Google emulator and how we can use it as a testing environment. Finally, I showed you how to get a performance score and rating for our website using Google PageSpeed Insights and YSlow. Using that report's recommendations, we can enhance our website's performance by making simple tweaks.

In the next chapter, I am going to discuss how to manage third-party components to increase our website's performance.

6
Managing Third-Party Components

In the previous chapter, we have discussed how to monitor and debug a mobile website using browser's DevTools. Most of the modern web browsers now include an advanced toolbar that helps developers to monitor and optimize their websites with ease. Many of these toolbars have the same features list, so if you could master one toolbar, I am sure you could apply the same mechanism for the others. Also, we have briefly discussed Google PageSpeed Insights and YSlow, which you can use to check your website's performance. The most valuable thing about these tools is that you can use their recommendations to enhance your website's performance.

In this chapter, I am going to discuss *how to manage third-party components* for optimal results. This is another crucial chapter for optimization, however, we tend to forget to check and apply many of following points. To go through this chapter, you don't have to be a pro, a basic knowledge of web development is sufficient.

I am going to discuss the following topics in this chapter:

- Eliminating 404 errors and missing assets
- HTTP 300, 400, and 500 codes
- **Content Delivery Network (CDN)**
- Third-party plugins
- Opening connection
- Closing connection
- Offloading to Wi-Fi
- Screen rotations
- Flash files

Eliminating 404 errors and missing assets

Getting 404 errors is the most annoying thing that the user can have when they visit your website. Search engine crawlers have also taken this up as a serious issue. If you use the Google analytics tool on your website, you can get all the 404 pages in your website. To do this, navigate to **Google analytic | Behavior | Site Content | Content Drilldown** and search for 404.

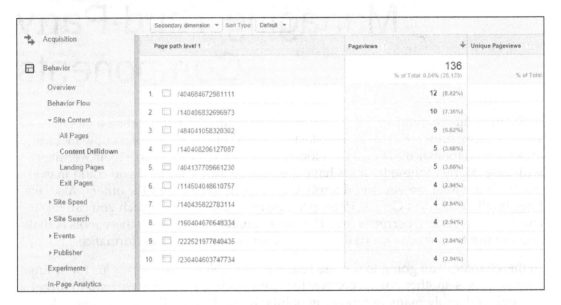

Using this information, you can fix your site's 404 errors with ease. Also, remember that sometimes we can't prevent the occurrence of all the 404 errors on the website because we are constantly working on our website, and it's a never ending process. However, make sure that you are aware of it so that you can fix it later.

Also, you can find the 404 errors that the search engine crawlers have encountered using webmaster tools. There are three main webmasters tools available out there and they will tell you which 404 errors they have found:

- Google search console
- Bing web master console
- Yandex web master

When you have found a 404 error in your website, and you know the reason for its occurrence, you should fix it as soon as possible. This will indicate the quality of your website for both users and search engines.

However, although you may have fixed these 404 errors on your website, it will take a while for the search engine to index them. So, if you can redirect those pages to appropriate pages, then do so.

To create redirects, you can use the following methods:

- **Create them manually in the .htaccess file**: This is the fastest way that you can follow if you know how to access the .htaccess file.

- **Use a redirect plugin**: There are a number of redirect plugins available in the market; you can use any one of them depending on your platform. However, use this with caution because sometimes these plugins will slow down your application.

Also, going through your server logs, you can get a different kind of 404 error other than missing pages, such as broken images or broken videos. Additionally, beware of broken embedded video URLs such as YouTube because sometimes they do not appear in server logs. All of these missing assets don't cause the whole page not to work, but they do look sloppy. These types of issues are more difficult to find because web master tools don't report them correctly.

The easiest method to find these missing assets is to use a tool such as **Screaming Frog**. The tool is very reliable when it comes to finding broken images. Also, you can go through your server log and do a search using a combination such as 404 and .jpg/.png.

I suggest, as a practice, that you check your website's 404 errors every month. If it's a large website that has a massive number of pages, then checking for errors every week is recommended. When you run the test at first, you will get a huge numbers of 404 errors but gradually, this number will decrease.

HTTP 300, 400, and 500 codes

If you get a 3XX class of HTTP status code, it means that the user agent needs to take some additional actions to fulfill the request. The required action is fulfilled by the user agent without any involvement of the user, if and only if, the method used in the second request is HEAD or GET. A client needs to detect infinite redirection loops in the website because for each redirection, it will generate network traffic.

- **Receiving an HTTP 301 status code**: An HTTP 301 status code specifies that requested resources by the user have permanently moved to another location. The HTTP response that carries this code needs to also include the new location, and the client should use this new URL the next time they try to get the same resource. If it's possible, a client needs to update all references to the requested URL when an HTTP 301 status code pops up.

- **Receiving an HTTP 302 status code**: An HTTP 302 status code specifies that requested resources by the user have temporarily moved to another location. The HTTP response that carries this code needs to also include the new location. This code implies to the client that they need to continue using the same URL to access this resource.

- **Receiving an HTTP 303 status code**: This status code specifies that the server sent this response to direct the client to get the requested resource to another URL with a GET request. It's essential to know what the HTTP 400 and 500 error codes mean so that you can fix the issues that they raise, especially if you have control over the cause.

- **Receiving an HTTP 400 status code (bad request)**: This error specifies that the user's request carries incorrect syntax.

- **Receiving an HTTP 401 status code (unauthorized)**: This error specifies that the requested file needs authentication (a username and password).

- **Receiving an HTTP 403 status code (forbidden)**: This error specifies that the server is not allowed to access the requested file. If a user received this message, you need to check the file permission settings or verify that the file is protected.

- **Receiving an HTTP 404 status code (not found)**: We have already discussed this code in a previous section. If you are getting a 5XX error, it means that your server cannot fulfill a valid request from a visitor. To resolve this type of issues, most of the time, you need to get support from a server administrator.

- **Receiving an HTTP 500 status code (internal server error)**: Usually, you get this type of error code because of incorrect server configuration.

- **Receiving an HTTP 501 status code (not implemented)**: This error occurs when the user's request is not supported by the server. This can happen because of outdated servers.

- **Receiving an HTTP 502 status code (bad gateway)**: This error occurs because of the incorrect configurations of proxy servers. Also, the error could arise because of poor IP communication between backend computers.

- **Receiving an HTTP 503 status code (service unavailable)**: This error occurs when the server has been temporarily closed due to maintenance or because of a large number of requests.

- **Receiving an HTTP 504 status code (gateway timeout)**: This error occurs when the server doesn't get a response from a server further up the chain. This issue happens mainly because of slow communication between upstream computers.

- **Receiving an HTTP 505 status code (version not supported)**: This error message occurs when the server refuses to fulfill the request made by the client. This issue happens mainly because of the invalid protocol specified by the customer device.

- **Receiving an HTTP 506 status code (variant also negotiates)**: This error specifies wrong server configurations.

- **Receiving an HTTP 507 status code (insufficient storage)**: This error occurs because of memory outage of the server.

- **Receiving an HTTP 509 status code (bandwidth limit exceeded)**: This error occurs when the bandwidth limit exceeds the limit that was set by the system administrator.

- **Receiving an HTTP 510 status code (not extended)**: This error occurs when an extension attached to an HTTP request is not supported by the server. As I have explained, every developer needs to know the meanings of these error codes. Then, they can understand the root cause and act upon it. Not only for mobiles, but this is also crucial for every platform.

Content Delivery Network

The user's geographic location and the distance on which your web server has a direct impact on response time is limited. So, dividing your website's content across multiple geographic locations will enhance page loading time. A **Content Delivery Network (CDN)** hosts files in different locations so that the person who visits your website can receive the nearest copy faster.

Most CDNs are used to host statistic content such as JavaScripts, CSS, images, videos, and fonts. Other than these, there are many benefits that we can get if we use a CDN network:

- **Different domains**: For a single domain, there is a limit for the concurrent connection made by browsers. Most of the time, it's between 4-8 and other connections have to wait until they are completed. However, CDN files are hosted on a different domain, so the browser permits to download CDN hosted files exceeding the normal limit.

- **Files can be pre-cached**: There is a high probability that someone may visit the website using Google CDN before you visit (files such as jQuery). So, files can be already cached, and you don't have to download them again.

- **High-capacity infrastructures**: You can get a flexible and reliable service from a reputed CDN provider who will have a higher availability, lower network latency, and less packet loss.

- **Distributed data centers**: If your web server is located in Texas and the user is visiting from Asia, they have to make a number of transcontinental electronic hops before they can access your files. Using a CDN, we can eliminate this because it serves the closest data to the user.

- **Built-in version control**: Most of the CDN has an integrated version control, so you can serve your desired version to the end user.

- **Boosts performance and saves money**: Using a CDN, you can distribute the load, increase performance, and save bandwidth, which is the expected goal when it comes to mobile web optimization.

Third-party plugins

Connecting and using third-party applications has increased in the last couple of years, and this may slow down your application. I think this is a huge issue that we face these days, and if an external plugin takes a longer period to load, it will have an enormous effect on customer experience.

Most of the time, third-party applications provide a broad range of functionality, such as widgets, analytics, ads, and tracking software. However, they can cause performance issues.

The issue with third-party applications is responsiveness and availability. Some scripts may be optimized for performance but most are not. Also, even if you are using a reliable third-party script, it may have an outage.

Using a third-party application may bring many functionalities to your application. However, it is important to decide if it is worth the risk because one line of third-party application code can crash your application.

To minimize the impact of the third-party application on your application, consider the following recommendations:

- Weigh the benefits versus risk value on performance and use add-ons as much as possible

- If you have decided to use third-party applications, load them asynchronously

Opening connection

When you load a JavaScript file synchronously in the Head tag of an HTML page, other requested files have to wait until the JavaScript file gets downloaded. This is more visible to the end-user because, he cannot see any content until all the files in the HEAD area get downloaded. However, using the asynchronous method to download the file we can resolve this issue to some extent.

When we specify JavaScript files to be loaded in the HEAD section of the HTML, it is essential to know the two different ways that we can use to load the data, which are as follows:

- **Synchronous**: The requested file will load before the parsing of the page continues

- **Asynchronously**: The other requested files will be downloaded parallel to the requested file

So, the best practice will be the JavaScript files located in the HEAD section of the HTML file load asynchronously. By doing so, we can eliminate page rendering delays.

Closing connection

Most of the time, many mobile applications leave their connections open after they have transmitted necessary data. This usually happens if we don't close the connection manually, once our transmission is completed.

As a practice, you need to always close the opened connection as soon as possible once the transfer is completed. Promptly closing connections eliminates you from having to open the purpose of closing connection due to timeouts. By discarding as many of these inefficient connections as you can will help you to save energy and optimize your application.

Offloading to Wi-Fi

If you are connecting to the network using 3G, 4G or HSPA regardless of the carrier, they use radio state machine to manage the radio resources. On the other hand, Wi-Fi connections are much more efficient than 3G and 4G. For Wi-Fi, there is no need for the state machine. As a developer, you can use this to your advantage.

Although there is less latency for connection setup, many applications fail to obtain this advantage. Offloading to Wi-Fi offers has several benefits, which are as follows:

- Lower data cost
- Extended battery life
- Faster connection
- Improved network traffic
- Improved customer experience

However, if you use Wi-Fi extensively, it will drain your battery faster than compared to a 3G or 4G data connection. However, if you use it occasionally, the impact will be minimal.

You can use a Wi-Fi connection for the following situations:

- If your application requires real-time interaction with the server, Wi-Fi will be the best option.
- If you have a heavy data usage application, use Wi-Fi as much as possible. You can use a tool such as WifiManager to scan for Wi-Fi networks near you.

Screen rotations

The introduction of the accelerometer in mobile devices has created new opportunities for developers to create an application that is creative, innovative, and user-friendly. The accelerometer allows the device to detect the physical orientation of the device so we can use that information in our application. For example, if you open the calculator application on your device, in vertical view it will behave like a regular calculator. If you change the orientation to horizontal, the calculator will turn into a scientific calculator. However, if we don't manage the screen rotation efficiently, it may cause issues.

With some mobile applications, screen orientation initiates network connection just for that event. This wastes the device's battery power and consumes network resources.

However, to use the accelerometer efficiently, you need to track the orientation locally and send these details at scheduled intervals, bundled with another event. Using this method, we can save power and data usage.

When we rotate our screen, we will get a different option and user interface:

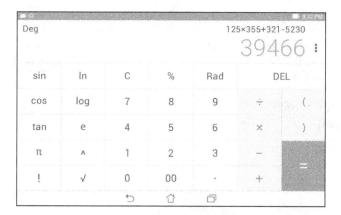

Flash files

Adobe Flash is used to create rich Internet applications, vector graphics, animations, and games. It's been very successful in the past but now the current versions of Android and iOS no longer support it. So, if your application uses Flash, your application will face a breakdown.

Because of the lack of support from the device's OS, using Flash content is not recommended:

- iOS don't support flash
- The Adobe Flash Player can only be supported on Android 2.2 through Android 4.0, Blackberry PlayBook, and HP webOS
- Adobe itself has stopped supporting the Flash Player

So the recommendation is that you need to avoid using Flash content on your website or application.

Summary

In this chapter, we have discussed how we can check 404 errors in our website, why it is important to eliminate 404 errors, and how we can do so. Not only 404 errors, but we also learned and understood 300, 400, and 500 error messages as well. Then, we have discussed CDN networks and the benefits of using a CDN network. Also, you need to manage third-party plugins in your website properly in order to get an optimized website.

In the later part of this chapter, I have explained how the opening and closing of a connection works and the importance of offloading to Wi-Fi. After this, we have discussed screen rotation and how we can use it to optimize our website. Finally, we went through Adobe Flash and recommended not use it.

In the next chapter, which is also the final chapter, I am going to give you some tips and tricks that you can use to enhance your application's performance.

7
Tips and Tricks

In the previous chapter, we discussed the reasons why we should eliminate 404 errors and missing elements on our mobile website. Also, the developer should have an excellent knowledge of 300, 400, and 500 codes, and I have briefly explained these code types. Using a CDN is a superb method to boost your website, and it allows you to deliver your website content faster to the end-user. Managing third-party plugins in the proper way is really essential because it can make or break your website. I have explained the importance of opening and closing the connection and managing it properly; you can enhance the performance of your website as well as save bandwidth and battery power of the user's device.

We have discussed the benefits of offloading through Wi-Fi rather than using 3G, 4G, or HSPA for connectivity. Finally, I have explained how to manage screen rotation and why we should eliminate Flash file usage on mobile devices.

With the information given in all the chapters that you have gone through so far, I assume you have learned many mobile optimization techniques. It's true that when it comes to the real world, applying all of these techniques to a website is not possible. However, as a developer, manager, or a business person you should try to implement these as much as possible.

In this chapter, which is also the last chapter, I am going to discuss a few tips and tricks that we can use to enhance mobile web performance. The topics covered are as follows:

- Built for performance
- When to optimize
- Invest for performance
- Design tool
- Performing actions optimistically

- Move bits when no one is watching
- Less work for the end user
- New relic

Built for performance

Now we all know why performance optimization is important for a mobile application or website. However, to implement and develop websites for performance, it takes time and money. If you fail to get budget approval from stakeholders for a project before it begins, you will have a really hard time. However, convincing them to do so is not that easy because usually, no one wants to invest if there is no valuable return to them.

It's easy to get excited about reducing metrics like load time and page weight, but they're probably not what matters to the people you need to get support from. Most of the time, stakeholders want to hear about what optimization will do for the things they care about. Some people will want to observe how it affects the bottom line. Others may be concerned more about what it means for page views and bounce rates. Learn what others care about and focus on prioritizing how performance improves those factors. You'll have a lot more success convincing them of the importance of performance if you can show some statistical data that matters to them.

If you are pitching your idea and importance of performance to a client or top management, use visuals to convince them. One excellent method is to show the client their competitor's application or website's performance and convince them how to take advantage over their client's website. If your customer is smart enough, they will understand this straight away because no one likes to lose to their opponent.

There are a couple of tools available online and by visiting `www.webpagetest.org` you can do this easily. Once you load `www.webpagetest.org`, go to **Visual Comparison**.

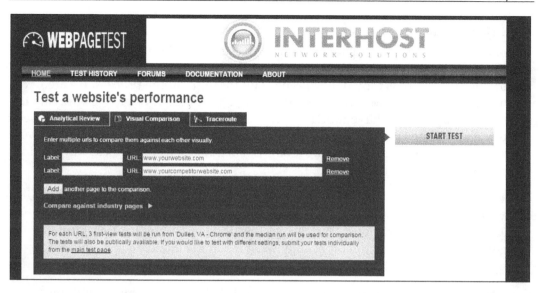

Type in the URLs that you want to compare:

Once you click on the **START TEST** button, the tool will begin to generate a comprehensive report. Using the **Create video** button you can make a video as well. This tool will capture screenshots throughout the loading process so you can see exactly when things start to load on each second.

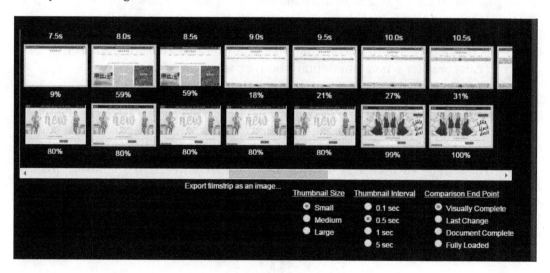

When to optimize

When it comes to web optimization, most of us leave the discussion of performance to the end of the conversation, when we mention it in passing; we underplay its importance to the project. By not bringing it up throughout the process, we are saying that we don't think it is important enough to discuss further. We're saying it's something that does not have much value. If we want to start correcting the course of performance on the Web, we must have the discussion on performance from the very beginning of the process, and we must be strict about it. One of the best ways to do this is to set a performance budget.

Most of the time we hear many people say in the early stages of a project that they want their site to be fast, only to see it turn into another one of those things that would be nice to fix eventually. We can get this done by setting up a performance budget. A performance budget is exactly what it sounds like. You set a budget for your page and do not allow the page to exceed that number.

Starting with page load time is a splendid idea, but the budget you set and refer to will hold more weight if you can specify the actual page weight or request count. Referencing a particular page weight or a number of requests instead of just a particular load time makes the conversation easier.

For example, suppose your budget says that the site must load in less than six seconds on a 3G network, and you're trying to decide whether or not to add a gallery to the page. To do this, you must first translate those six seconds into a weight or request count to be able to make that determination upfront.

Request count and page weight are also a relatively easy thing to reinforce in your build process, allowing you to enforce the budget rigidly if you so choose. Knowing that performance affects just about every important business metric, the ideal scenario is to make your site as fast as possible. Also, get to know about as many details as possible is critical to web optimization. There are two additional criteria that you can consider when you define a budget, which are as follows:

- **Current performance of your website**: First of all you have to audit and measure your website's performance under different network conditions. When you do that test, make sure that you measure the loading time, page weight, and the number of HTTP requests that you made.

- **Current performance of your competitor's website**: Do the same test for other similar sites in the industry.

Then, you can get an idea of how your websites perform and using this data you can decide your budget.

Also, when you define a performance budget, be realistic. Make sure that you set goals that you can achieve, such as *No more than 20 HTTP requests* or *maximum page size going to be 600 KB*. If you target an unrealistic budget, chances are you are not going to achieve it. So, be strict, but understand the reality.

Invest for performance

Performance budget is an excellent way to ensure performance remains part of the project. However, as we discussed earlier, the budget must be defined as soon as possible. If you are half way through a project without a performance budget, you will have a difficult time to convince the stakeholders or top management. Also, by then, there may already be approved designs or features that immediately crush whatever budget you may have needed to set.

Another important thing is that when you define the budget, and as you enforce it, you should already know what content needs to be on the page. When you allocate a performance budget to your project, it should help you to decide how to display your content, rather than not what content to display. Removing content from your page is not a strategy of performance.

Design tools

Mobile web optimization requires teamwork; it's not just the developer's job to enhance the performance of a website. Project plan, designs, developments, and QA all should be linked together to achieve a performance goal. If we take the designers, they play a huge role in developing a website. So, they have to be involved when they do their visuals. For example, we all know how flawed designing websites with software like Photoshop is.

I think the more it gets discussed, the more we realize that, as with any tool, Photoshop has many positives as well as negatives. We'll never fully get rid of image manipulation software in the design process and, frankly, that shouldn't be the goal. But it's important to consider what shortcomings it has so that our process can take them into account.

One valid concern about spending too much time in Photoshop versus the browser is that you see a picture of a website under ideal situations and at a particular, exact size. This is frequently cited as a problem for responsive design, but it is also a performance issue. Usually you can't mock-up performance in Photoshop.

Checking your application or websites in a browser beforehand can help you identify potential performance bumps before they have a chance to get completely out of control. That mock-up where every element has a semi-transparent shadow may look beautiful, but when loading it on a mobile device you may notice that scrolling is an arduous task. Identifying the problem at an early stage allows you to consider other solutions.

To be clear, these other solutions needn't be devoid of those kinds of embellishments altogether. Performance sites needn't be visually ugly or boring. There's a series of trade-offs to be made by weighing the benefits and the costs. Performance and visual creativity are both necessary, your site needs to balance the two. Getting real code on real devices as early as possible will help you to maintain that balance.

One of the best ways to allow you to get into the browser early is to think about your site regarding reusable components. Have a question about how that fancy navigation embellishment is going to perform on different devices? Load that in the browser, develop that part, and take it for a test run.

There are a number of ways to do this. My recommendation is to break website elements down into their smallest forms allowing you to build, say, a footnote, without committing to making the rest of the page as well. This enables you to quickly see how different pieces may work at different resolutions and with different input methods. The particular tool you use is less important than the result: being able to quickly test bits and pieces to identify performance issues before you're too far down the road to turn back.

Performing actions optimistically

If one of your visitors on your website decided to leave a comment, they would click on a button that submits the form. When they do that, two actions take place. The form, using AJAX, sends a request to the server, and a loading graphic appears to notify the user that their submission is in the process. When the script hears back from the server that the task has been completed successfully, it updates the page alerting the visitor.

This is the way it's usually done, but maybe it's not the best way. This request, particularly on a high-latency network, can take several hundred milliseconds, which is a very noticeable delay for the person trying to submit their comment.

Instagram has taken a different approach to avoid that delay. As soon as the person submits the comment, it appears on the page. The request happens in the background. To the person submitting the comment, it looks like it happens instantaneously. In reality, it takes as long to process as any other form online but the perception is dramatically improved.

This is called asynchronous UI, and we have discussed that in a previous chapter. The idea we use here is the same. We removed the loading state and let the user feel that things are moving more quickly. If the task fails, then gently alert them somehow and let them resubmit immediately.

Another great example is Polar (`https://www.polarb.com/`), the popular polling application. When you create a poll, it shows up instantly on your feed. Again, there's some smart asynchronous UI at work. What actually happens when you create a new poll is that Polar creates a temporary local copy of the poll and pushes it to the top of your feed. The temporary copy is fully functional. You can vote and comment on it, and those votes and comments will get pushed to the actual poll once it's been uploaded.

In the background, Polar uploads the temporary copy to its servers. If that fails, they try again a few times before finally admitting defeat to the user. The result, once again, is that the process seems to appear as incredibly fast.

It's important to note that for both Instagram and Polar, these solutions are not exactly ideal from an engineering perspective. There's a hell of a bit more complexity involved. But the trade-off is that the users get a system that feels instantaneous.

Move bits when no one is watching

When users sign up for Instagram, they are asked to fill out some basic details. While this is going on, in the background, Instagram starts looking for recommendations on who to follow.

The result is that by the time the user submits the form with their account information, recommendations are presented nearly instantly.

Instagram uses the same trick on image uploads. After you select the filter for your image, you can choose options such as where to share the image, or geotag it with a location. All the while, Instagram is already uploading the image in the background to reduce the time users have to wait at the end of the process.

It is worth noting that in both of the cases, there's a very high likelihood that the person will end up moving forward to those next pages. It's a slippery slope between moving bits while no one is watching and using up everyone's bandwidth and data for pages they may never view and assets they may never need. But if there is a high likelihood that your visitor will end up needing these assets at some point, it makes sense to do a little pre-caching to stay a step ahead.

Less work for the end user

Each of the principles mentioned earlier is ultimately about reducing the amount of time it takes for the user to complete a given task. The importance of task completion can't be overlooked. There's the classic study conducted by UIE back in 2001 on the impact of the time taken to complete a task on a visitor's perception of performance. Researchers sat people down in front of 10 different sites using a 56 kbps modem and gave them tasks to complete.

The surprise came when people rated the slowest site (http://www.amazon.com/) as one of the fastest when asked. The reason was that http://www.amazon.com/ allowed people to complete their tasks in fewer steps.

Ultimately, this is what it comes down to: how fast the user feels the site is. You can go a long way by implementing the performance techniques so frequently cited for developers, but to influence how your users feel about the performance of your site, performance optimization has to involve the designer.

If you're a designer, consider yourself the first line of defense. Yes, ultimately the developer will have to make many of the specific optimizations, but you are the person who gets to set the stage. You must make the decisions early on that will either encourage the site to be as fast as it is beautiful or encourage it to be beautiful, yet bloated.

New Relic

In *Chapter 5, Monitoring and Debugging Our Website*, we have discussed a few tools that we can use to debug our mobile application or a website. Using those tools you can get an idea about your product and how it behaves. However, using New Relic Mobile you can get a comprehensive idea about your project, and it helps you to build high-performance, stable mobile applications with ease.

The most challenging part of application building is that we can't predict how it's going to behave in the complex real world environment. New Relic's **Mobile APM (mAPM)** toolset fetches performance data to the developers, so they can focus on ensuring that users undergo a great experience. With New Relic Mobile, you can get the following features:

- **See how services are affecting your app**: Quickly find and fix performance issues caused by third-party API calls or internal backend services
- **Diagnose performance by operating systems, devices, and versions**: View and categorize performance reports into a particular one across all OSs, devices and app versions
- **Discover regions affected by poor app performance**: Observe the effect of your app's performance across geographical regions because of the response time, network failures, and data transfer
- **Understand how your app runs on a different carrier's network**: Get your app's breakdown of performance across each wireless carrier or Wi-Fi network

Following areas will check by New Relic Mobile APM

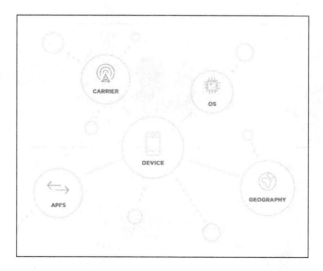

Also, it is essential to learn how the performance issue affects the user's experience. With New Relic Mobile, you can see the code-level inefficiencies related to how your customers use your application. In other words, it helps you to see your application's behavior as the end user. Using this tool you can achieve the following:

- **Code-level visibility for a user's interactions on your app**: View detailed timeline breakdowns of your slowest or most expensive interactions according to data distribution across background and foreground threads.

- **Identify how your apps consume device resources**: Get an idea about your application's CPU and memory usage at the local device level.

- **Track unique interactions with custom events**: Monitor a custom event in the application as interactions, business transactions, and in-app purchases. New Relic provides the same degree of granularity for all user events.

Crash reporting is the newly introduced feature by the New Relic Mobile. It helps a user to identify the reasons behind app crashing and helps to fix them immediately and efficiently, and prevents it from happening again.

- **Minimal effort, smarter troubleshooting**: The tool identifies the crashing patterns and rolls up the important details.

- **Understand how and why your apps crash**: The developer can determine when the application crashes by observing recurrence patterns and stack follows. Once the developer has fixed it, the tool utilizes week by week correlations to check whether the developer is enhancing the circumstance.

- **New Relic Mobile keeps you informed, so you never miss a problem**: With e-mail alerts, you will get a message when a crash happen, so you can fix it immediately.

We can't predict the future, nevertheless embrace it

We live in a rapidly changing and fascinating world. Every day, new devices, browsers, and technologies keep emerging. It's hard to keep a track on all the scenarios and build applications compatible with it. For example, building applications for screen size is a never ending process. At some point, we have to come to a conclusion as to what really matters.

We develop our process and design phases to the final product to keep our clients happy and give an ultimate experience to them so that they can do their task as efficiently and quickly as possible.

We are here to solve a problem and give a better solution to our customers. We must understand that, and we should see our solutions from their perspective. With this in mind, we can build an amazing experience for our clients. Finally, whatever the things you get as input, you should observe the situation, analyze it, and give an optimized solution to your customers.

Summary

In this chapter, we have discussed why we should build for performance and how we can convince our clients to approve a budget for performance. It is crucial to get the client on your side as early as possible. As I have mentioned, you can use an online tool such as WebPagetest to generate visuals so that you can convince your client easily. Sometimes, we tend to leave the optimization bit to a latter part of the project. This is really dangerous and ineffective, and you should plan and start optimization at the beginning of the project. Again, it is vital to invest in performance, and I have explained how we can do that aligning with your budget.

Also in the chapter, I have explained what the limitations in our design tools are and how we can get the best out of it. It's really hard to compare a design tool like Adobe Photoshop and a web browser, so you should test your design as soon as possible on your browser. In the next part of the chapter, we discussed how we can enhance the user's experience by applying some advanced techniques. Many of the popular solutions such as Instagram are using these methods to give an optimized solution for their customers. So, why can't we use these methods? Also, as a solution provider you should always expect less work from your clients, that's how they behave, and that's what they expect. For the sake of a good provider-customer relationship, you must always leave less work for your visitors.

With this chapter I am going to conclude the book. In this book, I have tried to cover many areas related to mobile web optimization and most of them are relatively easy to implement. However, I recommend that you to go through with each and every chapter carefully and thoroughly because, you will get a clear idea about what to do and what not to do when you build a mobile application or a website. Also, you should try to practice the aforementioned tools as much as possible because by doing so, you will get a superfast and error free end-product.

Thank you!!

Index

Symbol

404 errors
eliminating 104, 105

A

actions
performing optimistically 118, 119
Adobe Flash 111
AngularJS 75
Apple iOS 14
Aurelia 75

B

bandwidth 9
BlackBerry 10 OS 15
Bootstrap 73
browsers
about 10
Chrome 11
Firefox 12
Internet Explorer 12
Opera mini 13
Safari 11
URL 10

C

cache-control 68
caching 68
Chrome 11

compression tools, images
ImageOptim 38
Kraken 39, 40
Tiny PNG 38
connection
closing 109
opening 109
Content Delivery Network
about 103-107
benefits 107
content prefetching 69
crash reporting 122
CSS
minifying 64
none, displaying 59, 60
online tools, URL 65
CSS3
about 55
Filter property 50
styling options, using 48
using 46
CSS/JavaScript frameworks
about 72
AngularJS 75
Aurelia 75
Bootstrap 73
Ember 75
jQuery 74
Knockout.js 76
reference link 72
Semantic-UI 74
Susy 74
Uikit 73

resizing, for image resolution
    correction 35, 36
size 35
Instagram 120
Integrated Development Environments
    (IDE) 65
Internet Explorer 12
ipSecurity
  URL 30

J

JavaScript
  minifying 64, 65
  online tools, URL 65
  optimizing 76
jQuery 74
jQuery mobile
  features 74

K

Knockout.js 76
Kraken
  URL 39

L

lazy loading
  advantage 76
LESS
  about 62
  mixins 64
  partials 63
  variables 62
link attributes 71
Long-Term Evolution (LTE) 6

M

media queries
  attributes 58
  used, for creating images 60, 61
  used, for creating videos 60, 61
  using 58, 59

Microsoft Windows Phone 8 15
missing assets
  eliminating 104, 105
mixins
  in LESS 63, 64
  in SASS 63
Mobile APM (mAPM) 121
mobile battery
  3G wireless state machine, working 5, 6
  4G LTE wireless state machine,
    working 6, 7
  about 4
  connections, closing 8
  connections, opening 8
mobile device
  expected features 3
  features, for optimization 46, 47
  history 1-3
  uses 3
mobile only website, versus responsive
    website
  about 18
  combined files 22-24
  CSS sprites 25-27
  domain protection 19
  duplicate scripts, removing 28
  future-ready 19
  Gzip compression, enabling 28-35
  Http request 20
  image maps 27
  link equity 19
  rendering experience 19
mobile OS
  about 13, 14
  Apple iOS 14
  BlackBerry 10 OS 15
  Google Android 14
  Microsoft Windows Phone 8 15
mobile UX
  about 42
  enhancing 42
  importance 42

Thank you for buying
Mobile Web Performance Optimization

About Packt Publishing

Packt, pronounced 'packed', published its first book, *Mastering phpMyAdmin for Effective MySQL Management*, in April 2004, and subsequently continued to specialize in publishing highly focused books on specific technologies and solutions.

Our books and publications share the experiences of your fellow IT professionals in adapting and customizing today's systems, applications, and frameworks. Our solution-based books give you the knowledge and power to customize the software and technologies you're using to get the job done. Packt books are more specific and less general than the IT books you have seen in the past. Our unique business model allows us to bring you more focused information, giving you more of what you need to know, and less of what you don't.

Packt is a modern yet unique publishing company that focuses on producing quality, cutting-edge books for communities of developers, administrators, and newbies alike. For more information, please visit our website at www.packtpub.com.

Writing for Packt

We welcome all inquiries from people who are interested in authoring. Book proposals should be sent to author@packtpub.com. If your book idea is still at an early stage and you would like to discuss it first before writing a formal book proposal, then please contact us; one of our commissioning editors will get in touch with you.

We're not just looking for published authors; if you have strong technical skills but no writing experience, our experienced editors can help you develop a writing career, or simply get some additional reward for your expertise.

Mobile Web Development

ISBN: 978-1-84719-343-8   Paperback: 236 pages

Building mobile websites, SMS and MMS messaging, mobile payments, and automated voice call systems with XHTML MP, WCSS, and mobile AJAX

1. Build mobile-friendly sites and applications.

2. Adapt presentation to different devices.

3. Build mobile front ends to server-side applications.

4. Use SMS and MMS and take mobile payments.

5   Make applications respond to voice and touchtone commands.

Mobile First Design with HTML5 and CSS3

ISBN: 978-1-84969-646-3   Paperback: 122 pages

Roll out rock-solid, responsive mobile first designs quickly and reliably

1. Make websites that will look great and be usable on almost any device that displays web pages.

2. Learn best practices for responsive design.

3. Discover how to make designs that will be lean and fast on small screens without sacrificing a tablet or desktop experience.

Please check **www.PacktPub.com** for information on our titles

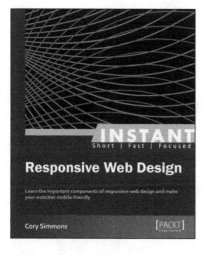
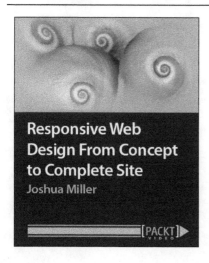

www.ingramcontent.com/pod-product-compliance
Lightning Source LLC
Chambersburg PA
CBHW060147060326
40690CB00018B/4011